NANOTECHNOLOGY SCIENCE AND TECHNOLOGY

THE NATIONAL NANOTECHNOLOGY INITIATIVE

FIFTH ASSESSMENT AND SUPPLEMENT TO THE PRESIDENT'S 2015 BUDGET

NANOTECHNOLOGY SCIENCE AND TECHNOLOGY

Additional books in this series can be found on Nova's website under the Series tab.

Additional e-books in this series can be found on Nova's website under the e-book tab.

NANOTECHNOLOGY SCIENCE AND TECHNOLOGY

THE NATIONAL NANOTECHNOLOGY INITIATIVE

FIFTH ASSESSMENT AND SUPPLEMENT TO THE PRESIDENT'S 2015 BUDGET

MARLIN THORNE
EDITOR

nova publishers
New York

For permission to use material from this book please contact us:
Telephone 631-231-7269; Fax 631-231-8175
Web Site: http://www.novapublishers.com

Additional color graphics may be available in the e-book version of this book.

Library of Congress Cataloging-in-Publication Data

ISBN: 978-1-63463-694-0

Published by Nova Science Publishers, Inc. † New York

CONTENTS

PREFACE

This book makes a number of recommendations in areas related to programmatic focus, program management, commercialization, measurements, and attention to environmental, health, and safety (EHS) concerns.

Chapter 1 – The National Nanotechnology Initiative (NNI) is a crosscutting national vision for nanotechnology development in the United States. The Federal effort in nanotechnology coordinates U.S. investment in research and development (R&D) in nanoscale science, engineering, technology, and related activities across the U.S. Government. In FY2014, even though five agencies garnered 93 percent of the Federal spending in nanotechnology R&D, 27 agency units from 20 top-level Federal entities participated in nanotechnology activities. The 21st Century Nanotechnology Research and Development Act of 2003 calls for a National Nanotechnology Advisory Panel to review the NNI periodically. Designated in 2004 to be that panel, the President's Council of Advisors on Science and Technology (PCAST) has reviewed the NNI five times, and this report is the third of this Administration's PCAST.

In the course of our overall assessment, we first reviewed the responses to the recommendations in the PCAST 2012 NNI review. One core recommendation was to increase funding to the Nanotechnology Signature Initiatives (NSIs). In contrast, we find that overall funding for the NSIs has remained flat, and we further find that funding for Solar Energy Collection and Conversion, Nanomanufacturing, and Nanoelectronics – the first three NSIs, which were created in 2011 – actually decreased; the proposed FY2015 NSI budget is down 28 percent from the FY2011 budget. The 2013 National Research Council Committee on the Triennial Review of the NNI examined the practices, agency collaborations, and progress of the 2011 NSIs and determined that only a few of the characteristics the Government Accountability Office (GAO) has published as best practices – purpose, scope, and methods; problem definition; and risk assessment – were addressed adequately; most of the initiatives only partially addressed other key characteristics such as resources, investments, integration, and implementation. The decrease in budget for the original NSIs and the NRC review imply that while NSIs may be good vehicles for identifying common areas in which participating agencies are investing, they are not serving as an effective vehicle for achieving the long-term vision of the NNI. Other 2012 recommendations included creating a standing Nanotechnology Steering Committee of outside experts and developing and tracking the metrics needed to quantify the Federal nanotechnology portfolio. Little activity has occurred that is responsive to these two important recommendations.

Chapter 2 – This document provides supplemental information to the President's 2015 Budget and serves as the Annual Report on the NNI called for in the 21st Century Nanotechnology Research and Development Act (P.L. 108-153, 15 USC §7501). The report also addresses the requirement for Department of Defense reporting on its nanotechnology investments, per 10 USC §2358. In particular, the report summarizes NNI programmatic activities for 2013 and 2014, as well as those currently planned for 2015. NNI budgets for 2013 through 2015 are presented by agency and PCA in Section 2 of this report. Information on the use of the Small Business Innovation Research (SBIR) and Small Business Technology Transfer Research (STTR) program funds to support nanotechnology research and commercialization activities, also called for in P.L. 108-153, is included at the end of Section 2. Section 3 outlines changes in balance of investments by PCA (including how agencies are allocating investments among the new PCAs outlined in the 2014 NNI Strategic Plan). Activities that have been undertaken and progress that has been made toward achieving the four goals set out in the NNI Strategic Plan (including activities in support of the NNI Nanotechnology Signature Initiatives) are outlined in Section 4. Highlights from external reviews of the NNI and how their recommendations are being addressed are summarized in Section 5. Appendices include supplementary information on nanotechnology investments by the Department of Defense, a list of abbreviations and acronyms, and a contact list of NSET Subcommittee participants and NNCO staff members.

In: The National Nanotechnology Initiative
Editor: Marlin Thorne
ISBN: 978-1-63463-694-0

Chapter 1

REPORT TO THE PRESIDENT AND CONGRESS ON THE FIFTH ASSESSMENT OF THE NATIONAL NANOTECHNOLOGY INITIATIVE*

President's Council of Advisors on Science and Technology

EXECUTIVE OFFICE OF THE PRESIDENT
PRESIDENT'S COUNCIL OF ADVISORS ON SCIENCE AND TECHNOLOGY
WASHINGTON, D.C. 20502

President Barack Obama
The White House
Washington, DC 20502

Dear Mr. President,

We are pleased to present to you the *Report to the President and Congress on the Fifth Assessment of the National Nanotechnology Initiative*, a review by the President's Council of Advisors on Science and Technology (PCAST). This report fulfills PCAST's responsibilities under the 21st Century Nanotechnology Research and Development Act (Public Law 108-153) and Executive Order 13349 to provide periodic updates to Congress.

The Federal Government has invested over $20 billion in nanotechnology research in the past 13 years, leading to great success in creating the building blocks of nanoscience. In this review, PCAST determined that the National Nanotechnology Initiative (NNI) has reached a turning point. The vision of NNI is a future in which the ability to understand and control matter at the nanoscale leads to a revolution in technology and industry that benefits society. To realize this vision, PCAST recommends that the Federal Government transition its activities toward facilitating commercialization by directing the formulation of specific nanotechnology Grand Challenges. The Grand Challenges framework—a partnership

* This is an edited, reformatted and augmented version of a report issued by the Executive Office of the President, October 2014.

between the public and private sectors— can drive scientific advances to revolutionary commercialized products.

PCAST recommends a process to identify the Grand Challenges and several program-management changes to ensure their success. New Federal activities can catalyze academic entrepreneurs and industry to come together and the manufacturing sector to consider producing these promising new technologies. The United States has made good progress in addressing environmental, health, and safety (EH&S) issues associated with nanotechnology, and the evaluation found that work on EH&S must continue so that new technologies are adopted with the full trust of the public.

Continued support of fundamental research in nanoscience is also critical. New technologies a decade from now will be built on exploration and discovery today. With strong awareness of how other countries are competing for the most talented scientists and engineers, PCAST recommends ways to attract and keep these individuals in the United States and sustain this country's advanced nanotechnology research infrastructure.

The transition toward commercialization can have implications for drug delivery, energy technology, smart sensors, clean water, quantum computing, and more. The United States can continue to lead in research and development, and the time is now to ensure the Nation will lead in the commercialization of nanotechnology, as well. PCAST thanks you for your interest in this important domain of American leadership in science, technology, and innovation.

Best regards,

John P. Holdren
Co-chair, PCAST

Eric S. Lander
Co-chair, PCAST

EXECUTIVE SUMMARY

The National Nanotechnology Initiative (NNI) is a crosscutting national vision for nanotechnology development in the United States. The Federal effort in nanotechnology coordinates U.S. investment in research and development (R&D) in nanoscale science, engineering, technology, and related activities across the U.S. Government. In FY2014, even though five agencies garnered 93 percent of the Federal spending in nanotechnology R&D, 27 agency units from 20 top-level Federal entities participated in nanotechnology activities. The 21st Century Nanotechnology Research and Development Act of 2003 calls for a National Nanotechnology Advisory Panel to review the NNI periodically[3]. Designated in 2004 to be that panel, the President's Council of Advisors on Science and Technology (PCAST) has reviewed the NNI five times, and this report is the third of this Administration's PCAST.

In the course of our overall assessment, we first reviewed the responses to the recommendations in the PCAST 2012 NNI review[4]. One core recommendation was to increase funding to the Nanotechnology Signature Initiatives (NSIs). In contrast, we find that overall funding for the NSIs has remained flat, and we further find that funding for Solar Energy Collection and Conversion, Nanomanufacturing, and Nanoelectronics – the first three NSIs, which were created in 2011 – actually decreased; the proposed FY2015 NSI budget is down 28 percent from the FY2011 budget. The 2013 National Research Council Committee on the Triennial Review of the NNI examined the practices, agency collaborations, and progress of the 2011 NSIs and determined that only a few of the characteristics the Government Accountability Office (GAO) has published as best practices – purpose, scope, and methods; problem definition; and risk assessment – were addressed adequately; most of the initiatives only partially addressed other key characteristics such as resources, investments, integration, and implementation[5,6]. The decrease in budget for the original NSIs and the NRC review imply that while NSIs may be good vehicles for identifying common areas in which participating agencies are investing, they are not serving as an effective vehicle for achieving the long-term vision of the NNI. Other 2012 recommendations included creating a standing Nanotechnology Steering Committee of outside experts and developing and tracking the metrics needed to quantify the Federal nanotechnology portfolio. Little activity has occurred that is responsive to these two important recommendations.

The Federal Government has proposed $1.537 billion of nanotechnology funding in FY2015, an amount roughly comparable to FY2014 funding. We find that the sustained level of funding to date has delivered significant scientific and technological progress and that continued innovation calls for a healthy research effort to continue. But international competition for leadership in nanotechnology has increased on many fronts, as evidenced by a 2014 GAO report stating that while the United States remains preeminent in nanotechnology research, it has fallen behind other countries in developing the necessary infrastructure and workforce required to manufacture many nano-based products[7].

Our work in this PCAST 2014 review is based on the premise that the nanotechnology field is at a critical transition point and has entered its second era, which we call NNI 2.0. This next technological generation will see the evolution from nanoscale components to interdisciplinary nano-systems and the movement from a foundational research-based initiative to one that also provides the necessary focus to ensure rapid commercialization of nanotechnology[8].

It is important to recognize that while Federal agencies prioritize nanotechnology as an area for investment and activity, the entire Federal activity is not a single, centrally-funded program with separate, line-item budget authority. The added value of the first phase of NNI in creating collaborations among various agencies is no longer enough given international competition and the maturation of the field. The primary conclusion of our 2014 PCAST review is that the United States will only be able to claim the rewards that come from investing in nanotechnology research and sustaining an overarching Federal initiative if the Federal interagency process, the Office of Science and Technology Policy (OSTP), and the agencies themselves transition their nanotechnology programmatic efforts beyond supporting and reporting on basic and applied research and toward building program, coordination, and leadership frameworks for translating the technologies into commercial products.

This report makes a number of recommendations in areas related to programmatic focus, program management, commercialization, measurements, and attention to environmental, health, and safety (EHS) concerns. Details are provided in the sections that follow. Here we note the three most important recommendations that we believe are necessary to bring about the focus and direction needed for NNI 2.0 to be successful.

1) While certain elements of the current Nanotechnology Signature Initiatives framework should be maintained, the primary active program-management structure should be driven by the Federal and OSTP commitment to the concept of nanotechnology Grand Challenges.
2) We reiterate the need for an ongoing, separate standing committee of cross-sector nanotechnology experts that advises, but does not evaluate, the nanotechnology activities of the U.S. Government. We also iterate the need for a functional interagency process via the National Science and Technology Council, the Committee on Technology, and the Nanoscale Science, Engineering, and Technology (NSET) Subcommittee that is able to make cross-agency funding priorities when needed to address nanotechnology Grand Challenges[9].
3) We reiterate the need to assess Federal nanotechnology research and commercialization funding through a more formal system of metrics.

Implementing the Grand-Challenges framework, continuing to support vital discovery and exploratory research, and continuing an active, collaborative, and competitive international engagement will ensure that NNI 2.0 is a successful and vibrant Initiative.

This report describes in detail the rationale and characteristics of Grand Challenges, how they differ from Signature Initiatives, and provides illustrative examples that could be suitable for NNI 2.0. Because the Grand-Challenges process outlines specific technical goals and the active management needed to accomplish them, that process, which has been used in a number of OSTP-led scientific and technology programs, can provide the necessary framework for commercialization opportunities to mature. Illustrative Grand Challenges for nanotechnology include nano-enabled desalination of seawater to solve the emerging water crisis, reducing global greenhouse emissions with nano-enabled solid-state refrigeration, creating a forefront of manufacturing through nano-3D printing, and developing a nanoscale therapeutic for at least one major cancer.

In addition to the three primary recommendations noted above (Grand Challenges, an active advisory committee, and development of processes to measure the effectiveness of the Initiative), this review makes additional recommendations aimed at enhancing the transition to NNI 2.0. We note that while these measures are meant to support the enhancement of nanotechnology investments that foster commercial transition of maturing nanotechnologies, this focus on commercialization should not preclude the Federal Government from investing in and coordinating discovery and exploratory research, which is the bedrock on which commercial technology builds. We recommend that the National Science Foundation (NSF) expand the NSF Innovation Corps to include a specific focus on entrepreneurship in the nanotechnology area; that with the guidance and support of NSET, the Federal agencies define potential Institutes for Manufacturing Innovation dedicated to nanoscience and nanotechnology as part of the National Network for Manufacturing Innovation program; and that Federal agencies extend or create programs that substantially support single investigators for five years to pursue creative, high-risk research in nanoscience and nanotechnology.

We commend NSET for its successful coordination of nanotechnology R&D activities within the Federal Government, and we specifically highlight the NSET Nanotechnology Environment and Health Implications (NEHI) Working Group. NEHI remains an active and collaborative cross-agency forum that released on June 26, 2014 a review of NNI EHS research activities, accomplishments, and collaborations entitled, "Progress Review on the

Coordinated Implementation of the NNI 2011 EHS Research Strategy." We recommend that NSET continue on this path and support further the development of a multidisciplinary nanotechnology EHS ecosystem that will expedite safety assessment, decision making, and commercialization. A fundamental Government role in attending to EHS issues will endure regardless of the shape of the Initiative.

The National Nanotechnology Initiative has been a truly successful venture for the past 13 years, and the nanotechnology community has built strong foundations for the future. NNI 2.0 will require an actively-led ecosystem that integrates the efforts of academic, industrial, Federal, and philanthropic partners and more to bring the vision of the NNI into commercial reality. Much of the analysis and many of the recommendations in this report are not new. With the enhanced focus on the transition to commercialization, the implementation of the Grand-Challenges framework, and more aggressive leadership, we believe the recommendations contained herein will lead to a successful NNI 2.0 for the coming decade. We believe the implementation of these recommendations to be critical to that success. If another two years passes without these kinds of changes, we believe the value of a centrally-reported NNI will be substantially decreased. The next PCAST review will most certainly focus on the success of this implementation.

SUMMARY OF RECOMMENDATIONS

Grand Challenges for NNI 2.0 Recommendations

Recommendation 1. Establishing Grand Challenges is an effective means for focusing and amplifying the impact of Federal nanotechnology activities. The Nanoscale Science, Engineering, and Technology Subcommittee and the Office of Science and Technology Policy should identify a list of candidate nanotechnology Grand Challenges that address significant societal needs. At least one Grand Challenge should contain program elements aimed at manufacturing challenges specific to that focus area.

Recommendation 2. The Nanoscale Science, Engineering, and Technology Subcommittee and the Office of Science and Technology Policy should create and execute a process to engage research, development, and industrial stakeholders in the identification and selection of Grand Challenges on an ongoing basis.

Recommendation 3. Federal agencies, with support from the Office of Science and Technology Policy, should offer implementation tools like innovation prizes and public-private partnerships to encourage researchers to reach critical milestones on the path to completing Grand Challenges.

Program Management Recommendations

Recommendation 4. The Nanoscale Science, Engineering, and Technology Subcommittee should continue to be co-chaired by the Assistant Director, Nanotechnology in the Office of Science and Technology Policy and a representative of one of the participating agencies on a rotating basis. The Assistant Director should ensure that a leader for each active

Grand Challenge also participates in a leadership role in the subcommittee. The Nanoscale Science, Engineering, and Technology Subcommittee should use Grand Challenges as the primary vehicle for actively managing the Federal nanotechnology activities toward directed outcomes.

Recommendation 5. PCAST should continue to carry out the Congressionally required periodic review of the NNI. Congress should align this review and the triennial National Research Council reviews to the same concurrent three-year time interval to reduce the burden on the Initiative. The Office of Science and Technology Policy, with the support of the National Nanotechnology Coordinating Office, should create and administer a separate standing committee of cross-sector nanotechnology experts that provides guidance, but does not evaluate, the NNI.

Recommendation 6. The Nanoscale Science, Engineering, and Technology Subcommittee, with the Department of Commerce, should execute a process to establish a common set of evaluation metrics to quantify and report the impact on workforce, productivity, and scientific knowledge in nanotechnology for all new research and commercialization programs beginning in FY2016.

Commercialization Recommendations

Recommendation 7. The Nanoscale Science, Engineering, and Technology Subcommittee should work with the Federal agencies to define potential Manufacturing Innovation Institutes dedicated to nanoscience and nanotechnology as part of the National Network for Manufacturing Innovation program.

Recommendation 8. The National Science Foundation (NSF), in consultation with the Nanoscale Science, Engineering, and Technology Subcommittee (NSET), should expand the NSF Innovation Corps to include a specific focus on entrepreneurship in the nanotechnology area, and NSET, the National Institutes of Health, and the Department of Energy should leverage this program concept to make it available to a broad range of scientists working in nanoscience and technology.

Recommendation 9. The National Nanotechnology Coordinating Office and the Department of Commerce should establish an annual nano-focused economic-development forum designed to bring together academic researchers, the venture capital community, biotechnology, and other industry in a format that enhances the possibility to create business partnerships.

Research Enterprise Recommendations

Recommendation 10. Recognizing growing international funding competition that is attracting US-based talent to go abroad, NNI agencies should substantially support the best single investigators to pursue creative, high-risk research. In particular, the National Science Foundation, Department of Energy, Department of Defense, and National Institutes of Health should coordinate to ensure that at least five new National Security Science and Engineering

Faculty Fellowship (NSSEFF)-style senior-investigator grants in nanoscience and nanotechnology are funded per year.

Recommendation 11. The National Science Foundation, National Institutes of Health, Department of Energy, Department of Defense, and the National Institute of Standards and Technology should strongly support nanoscale research centers and infrastructure networks to ensure the effective training of a new generation of transdisciplinary scientists and engineers, in particular by strongly supporting the Next-Generation National Nanotechnology Infrastructure Network.

Environmental, Health, and Safety Recommendations

Recommendation 12. The Nanoscale Science, Engineering, and Technology Subcommittee should continue to support the development of a multidisciplinary nanotechnology environmental, health, and safety ecosystem that promotes non-animal based (alternative) test strategies for safety assessment and multi-stakeholder participation in regulatory decision-making and safe implementation to facilitate market access of nanomaterials and nanotechnology-enabled products.

SECTION 1. INTRODUCTION & OVERVIEW

Introduction

This report conveys the fifth review of the National Nanotechnology Initiative by the President's Council of Advisors on Science and Technology (PCAST). To execute the review, we formed an 11-person working group of outside advisors who informed the evaluation.

PCAST strongly supports fundamental research in nanoscience and across the physical and nat- ural sciences. We found that Federal investment in nanotechnology has built a strong scientific foundation for creating, studying, and understanding new classes of nanoscale building blocks. We expect fundamental nanoscience research to continue uncovering new insights that bridge understanding between the atomic scale and the macroscopic scale. But the Federal Govern- ment's intense interest in nanoscale research and development is not driven by increasing fun- damental knowledge in nanoscale science for its own sake – a goal deeply appreciated by this Council – but instead aims to create new technologies that improve the lives of Americans and people everywhere. "There's plenty of room at the bottom," said physicist Richard Feynman about technology opportunities at the nanoscale. And this second goal drives the evaluation in this report.

The nanotechnology community has been growing since the 1980s. State governments now support university nanotechnology-user facilities, television programs and movies envision what a future with nanotechnology could look like, college students select majors in nanotech- nology rather than chemistry or materials engineering, and industry considers the tradeoffs in manufacturing nanotechnology and what producing nanomaterials means for workers' health. And since 2001 and the 21st Century Nanotechnology Research and

Development Act, the Fed- eral Government began to give a greater share of its research and development dollars to nan- otechnology research and development and a greater share of Federal effort to coordinate those investments.

The nanotechnology community is at a turning point. The research efforts of the last decade have delivered impressive understanding of the fundamental science relevant at the scale of individual nanoparticles and nanotubes. Efforts are now delivering equally important results as these "components" are assembled into nano- and meta-scale systems-product concepts. In parallel, the nanotechnology community's engagement with its stakeholders and the public has also matured. The community needs to consider what it wants to accomplish since the re- sources from universities, institutes, industrial facilities, Federal dollars, philanthropic founda- tions, and public interest are not infinite and come at an opportunity cost.

In light of the progress made and cognizant of budget pressures and potentially competing pri- orities for Federal investment, the NNI needs to have a clear, executable vision for how to max- imize the return on the substantial investments that have been made. Continuing to obtain new knowledge, developing new commercial devices, and improving processes or systems that enhance commercial transitions are critical steps to achieve that goal.

The promise of nanotechnology is great. It can potentially transform medicine, security, transportation, agriculture, and more, but for better or for worse, the future will also be shaped by competing technologies and approaches. During this transition, the nanotechnology community needs leaders who will articulate a vision for the community's future and who will develop strategic plans for what the community wants to accomplish.

We review in this report the actions and investments of one segment of this community: the Federal Government.

Federal Government's Role in Nanotechnology

The NNI vision is "a future in which the ability to understand and control matter at the nanoscale leads to a revolution in technology and industry that benefits society."[10] The Government's role so far in unlocking the potential of this technology has been to ensure that the activities within the U.S. Government are well coordinated. Since the NNI was launched in 2001, the Federal Government has brought together an increasing number of nanotechnology activities across its agencies. From FY2005 to FY2014 alone, 628 agency-to-agency collaborations resulted, growing from 35 such collaborations in FY2005 to 159 collaborations by FY2013 (Figure 1). Along with coordinating individual activities, in 2010 the Government's interagency process began to spotlight Nanotechnology Signature Initiatives (NSIs), which are rotating areas of national interest that involve investment and coordination of at least three Federal agencies. The collaborations around some of these NSIs have blossomed.

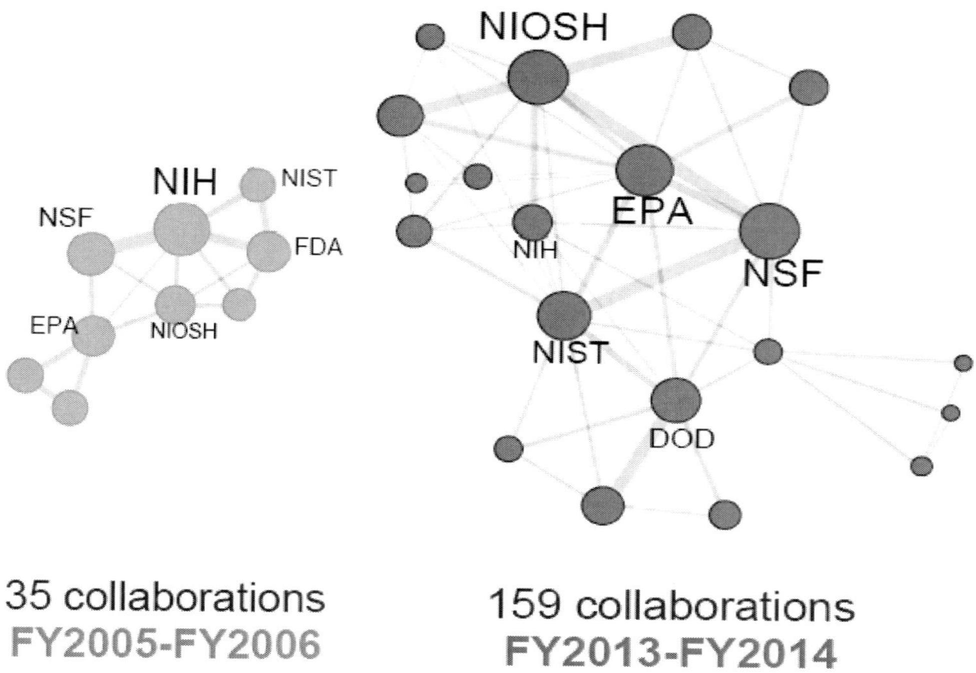

Figure 1. Federal interagency collaborations in NNI increased over four-fold in eight years.

After 13 years, the success of the first phase of activities and the maturation of the research field has placed the field of nanotechnology at a critical transition point. Success stories from just the past two years are evident in Boxes 1 and 2. We call the next phase of nanotechnology development NNI 2.0. In reviewing the Federal activities in nanotechnology, we sought to identify how best to invest Federal funds and to coordinate and lead Federal activities in the next decade. Strong management of program activities in commercialization, in the research enterprise, and in environmental, health, and safety issues remain as the core concepts. Whether our vision for NNI 2.0 is adopted or an incremental evolution of today's NNI is selected, the choice to have an initiative implies continuing to choose to make nanotechnology a distinct field of Federal Government focus.

Box 1. Nanotechnology Success Story #1

Light- and Heat-Shielding Nano-Coating for "Smart Glass" Developed at the Lawrence Berkeley National Lab

Researchers at the Lawrence Berkeley National Laboratory (LBNL) have developed an electrochromic composite nanomaterial that can be applied to glass to alter electronically the transmission of light and heat through the glass surface[1]. Electrochromic materials change color reversibly when a charge is applied. Electrochromic material-coated glass is currently marketed as "smart glass" and can transition between transparent and opaque states when a switch is flipped. Smart glass can improve temperature control or lighting systems, provide privacy, or act as a projection screen.

LBNL researchers embedded indium tin oxide nanocrystals in an electrochromic glass to develop a nanocrystal-in-glass composite. Properties of the nanocrystals keep certain infrared wavelengths from passing through the coating, and the glass material screens visible light. Using the two materials in a composite allows light and heat transmission to be controlled separately, and it makes the glass five times more electrochemically active so that researchers could use thinner and more transparent coats of the material. The system requires very little energy – a square-meter window coated in the composite material could be powered by a 9- volt battery for a year – making it much less expensive than current systems. It is also entirely solution-based and could be developed into a spray form. Researchers have launched a startup and are working to scale production and develop partnerships with the glass and building technologies industries.

Box 2. Nanotechnology Success Story #2

Nanotube-Infused Clothing May Protect Against Chemical Weapons

A research team at the National Institute of Standards and Technology (NIST) has completed a proof-of-concept demonstration of carbon nanotubes that degrade a class of chemical nerve agents and could be incorporated into fabrics to create protective clothing[2]. This class of nerve agents includes sarin, which was used in a 1995 bioterrorism attack in the Tokyo subway and more recently in the conflict in Syria. When absorbed into the body, these chemicals disrupt processes ranging from muscle contraction to high-level cognition and lead to long-term physiological and psychological damage and even asphyxiation at high exposures. In the NIST experiments, single-wall carbon nanotubes were bonded to a catalyst. Properties of the nanotubes improve the rate of catalysis to 63 times the untreated rate and maintain catalytic ability for several weeks. The nanotubes are flexible and lightweight with high tensile strength, making it likely that a fabric containing this material would be highly wearable in addition to self-decontaminating. This technology has the potential to protect both civilians and warfighters from chemical weapons and to improve the safety of cleanup operations. Researchers are working to increase the speed of the reaction and to determine the best method for incorporating the material in the fabric while maximizing its catalytic ability.

This introduction outlines the current organization of Federal activities under the National Nanotechnology Initiative and describes the state of international competition. In subsequent Sections, we evaluate and make recommendations on areas of program management; commercialization; research; and environmental, health, and safety (EHS) issues to realize the potential of this important field.

Organization of the National Nanotechnology Initiative

Numerous Federal agencies invest in nanoscience and nanotechnology. The President coordinates national science, technology, and space policy through the National Science and Technology Council (NSTC)[11]. NSTC's Committee on Technology has established its **Nanoscale Science, Engineering, and Technology** (NSET) Subcommittee to coordinate Federal activities with the National Nanotechnology Initiative vision. In addition to its formal relationship to the Committee on Technology, NSET also maintains a less formal relationship to NSTC's Committee on Science.

NSET membership consists of representatives from the more than 20 Federal agencies that have interests in nanotechnology, including more than 11 that are funding or have funded nanotechnology efforts over the history of the initiative, as well as representatives from the Office of Science and Technology Policy (OSTP) and the Office of Management and Budget (OMB)[12].

NSET currently charters two **Working Groups** to aid in its coordinating efforts[13]. These are:

> **The Nanotechnology Environmental & Health Implications Working Group (NEHI)** to provide a forum for focused interagency collaborations on EHS and leadership in establishing the national nanotechnology EHS research agenda, in addition to communicating EHS information amongst Federal agencies and to the public. NEHI participants include those agencies that are involved in policy, education, and science aspects of public, workplace, and environmental safety.
>
> **The Nanomanufacturing, Industry Liaison & Innovation Working Group (NILI)** to promote collaboration and partnerships across the broadest possible Federal, State, and private sectors to build U.S. leadership in the products and services that are and will be enabled by nanotechnology.

In addition to its Working Groups, NSET has appointed four **Coordinators** from across its members to strengthen interagency coordination on topics deemed critical to the success of nanotechnology development. Current Coordinators include the Coordinator for Standards Development; the Coordinator for Environmental, Health, and Safety Research; the Coordinator for Global Issues; and the Coordinator for Education, Engagement, and Societal Dimensions.

Central support to NSET as well as to the broader Federal investment in nanotechnology is provided by the **National Nanotechnology Coordination Office** (NNCO). NNCO was authorized in 2003[3]. It assists in public outreach, serves as a point of information exchange, and promotes access to foundational nanotechnology science and its emerging applications. The Office coordinates much of the public face of the Initiative, including staging meetings and workshops of the NSET and its working groups. NNCO coordinates preparation and publication of NNI interagency planning, budget, and assessment documents, and maintains the NNI website, http://www.nano.gov. NNCO is funded by NSET Subcommittee agency contributions[14]. Figure 2 shows the organization chart for the various participants in the NNI.

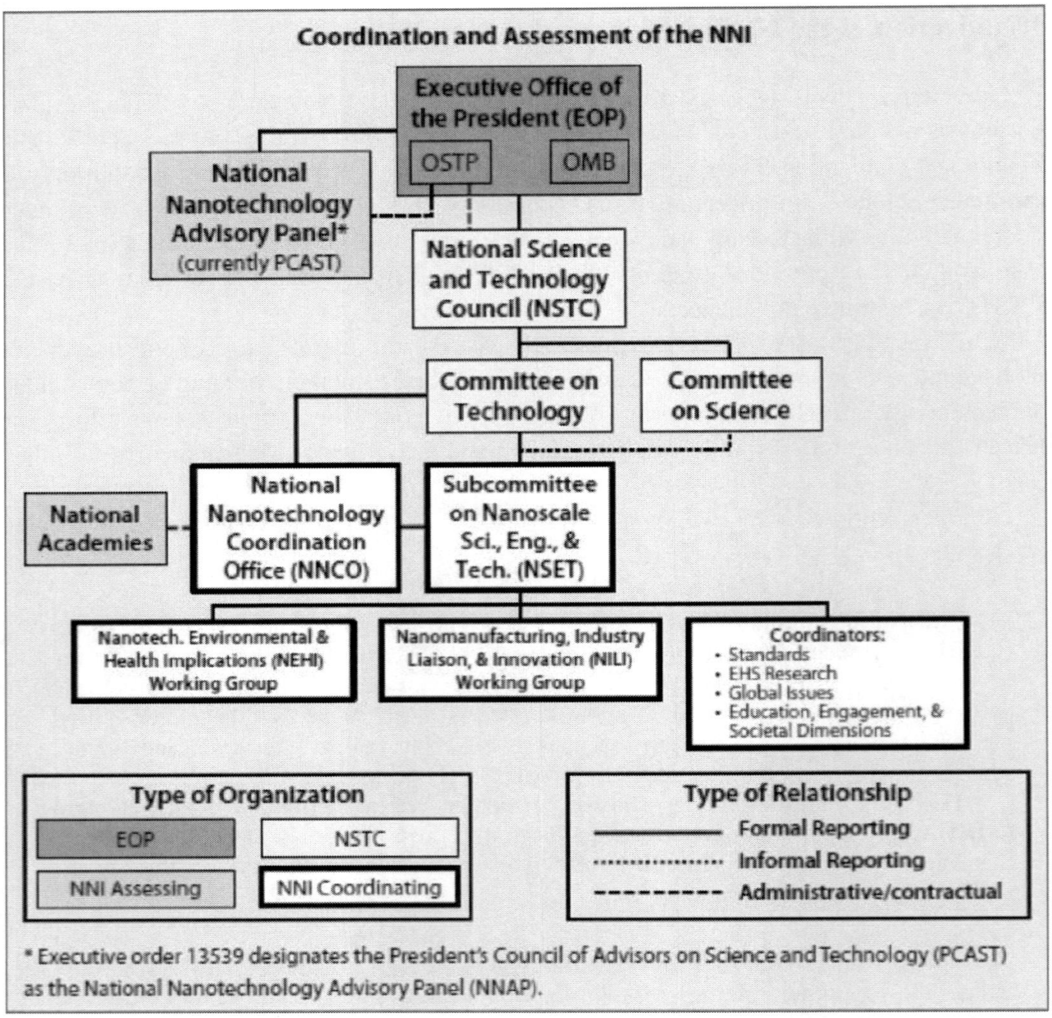

Figure 2. In the Federal structure committed to the vision of NNI, the Subcommittee on Nanoscale Science, Engineering, and Technology coordinates many interagency activities[15].

Funding the National Nanotechnology Initiative

Federal activities in nanotechnology range from investments in foundational research to efforts aimed at enabling commercial applications of nanotechnology-enabled products and solutions. Primary funding streams support individual and team-based research, a variety of multidisciplinary centers, and significant user facilities as part of the infrastructure necessary to support these research efforts. Activities relating to environmental, health, and safety concerns, as well as the development of policy frameworks to understand, educate, and guide the ethical, social, and legal implications of emerging nanotechnology, are also funded by Federal agencies.

Federal budgets for nanotechnology are developed by individual participating agencies and, as such, fit within the overall prioritization and approval process of those organizations, subject to review and approval by the Office of Management and Budget. The NNCO aggregates individual agency budget requests for nanotechnology and Congressional appropriations into a comprehensive annual report to provide visibility to the overall national investment in nanotechnology. The most recent report is the FY2015 NNI Budget Supplement[16].

Spending by agency has been reasonably stable over the last four years, with the notable exception of a 60% drop in Department of Defense (DOD) spending between 2012 and 2013, as sequestration and budget-reconciliation decisions were made. Historical nanotechnology funding by agency is detailed in Table 1, and the FY2015 request for funding by Program Component Areas is in Table 2[17]. As shown in Table 1, funding peaked for nanotechnology in 2010 at approximately $1.9 billion, dropped slightly to approximately $1.85 billion in both FY2011 and 2012, then dropped nearly 17% to the 2013 level. American Recovery and Reinvestment Act funding injected over $500 million in additional funds into the NNI, primarily at the Department of Energy (DOE), the National Institutes of Health (NIH), National Institute of Standards and Technology (NIST), and the National Science Foundation (NSF). Nanotechnology funding levels for 2014 are $1.537 billion; essentially flat to 2013's $1.550 billion on a non-inflation adjusted basis. As has been typical of recent years, five agencies account for over 93% of nanotechnology investments in the FY2015 budget proposal: NIH (28.7%), NSF (26.8%), DOE (22.3%), DOD (9.4%), and NIST (5.4%).

Overview of the Remainder of this Report

Evaluating the structure of the Federal investments in nanotechnology, the progress on the four goals of the NNI, the state of international collaboration and competition, and engaging in numerous conversations with members of the nanotechnology community brought up many ideas for PCAST to consider about the Federal Government's activities in nanotechnology. The primary conclusion of our review is that the United States will be able to claim the rewards that come from investing in nanotechnology research and an overarching Federal initiative only if the Federal interagency process, the Office of Science and Technology Policy, and the agencies themselves transition their nanotechnology programmatic efforts beyond primarily just supporting and coordinating basic and applied research toward building frameworks for translating the technologies into commercial products. To ensure that NNI 2.0 is a successful and vibrant initiative, we recommend the coordinated use of the Grand-Challenge framework as an overarching program-management structure, as has been effectively practiced in other parts of the U.S. Government research and development system. Grand Challenges are meant to align stakeholders across the nanotechnology community toward delivering transformative technologies. This Grand-Challenge framework in the U.S. Government must be supported by the development of an advisory committee to the NSET and the development of metrics to monitor the progress of Federal nanotechnology investment. We believe that by focusing on Grand Challenges, continuing to support vital discovery and exploratory research, and continuing an active collaborative and competitive international engagement, NNI 2.0 will deliver on the promises now emerging from the highly successful first phase of the National Nanotechnology Initiative. We note specifically that the focus on commercialization in NNI 2.0 should not preclude it from investing in and coordinating discovery and exploratory research, which is the bedrock on which commercial technology builds.

Table 1. NNI investments from FY2001 to FY2015 (request). Data provided by NNCO

National Nanotechnology Initiative Investments by Agency FY 2001-2015 (dollars in millions)																
FY	2001	2002	2003	2004	2005	2006	2007	2008	2009†	2010	2011	2012	2013	2014††	2015†††	Total*
CPSC	0	0	0	0	0	0	0	0	0.2	0.5	1.8	2.0	1.3	2.0	2.0	**9.7**
DHS	0	2	1	1	1	1.5	2.0	3.2	9.1	21.9	9.0	18.7	14.0	24.0	32.4	**140.7**
DOC / NIST	33	77	64	77	79	77.9	87.6	85.6	93.4	114.7	95.9	95.4	91.4	97.8	82.6	**1252.3**
DOD	125	224	220	291	352	423.9	450.2	460.4	459.0	439.6	425.3	426.1	170.1	175.9	144	**4786.4**
DOE	88	89	134	202	208	231	236	244.7	332.6	373.8	346.2	313.8	314.2	303.3	343.1	**3759.7**
DOJ	1	1	1	2	2	0.3	1.7	0.1	1.2	0.2	0.0	0.0	0.0	0.0	0.0	**10.5**
DOT	0	0	0	0	0	0.9	0.9	0.9	0.9	3.2	1.0	1.0	2.4	2.0	1.5	**14.7**
EPA	5	6	5	5	7	4.5	7.6	12.1	11.6	17.7	17.4	17.5	14.6	15.5	16.8	**163.3**
DHHS (total)	40	59	78	106	168	195.4	222.7	311.4	356.0	472.6	428.543	479.6	485.4	469.5	469.6	**4341.7**
FDA	0	0	0	0	0	0	0	0	6.5	7.3	9.9	13.6	16.1	17.0	17.0	87.4
NIH	40	59	78	106	165	191.6	215.4	304.5	342.8	456.8	408.6	456.0	458.8	441.5	441.5	4165.4
NIOSH	0	0	0	0	3	3.8	7.3	6.9	6.7	8.5	10.0	10.0	10.5	11.0	11.1	88.8
NASA	22	35	36	47	45	50	19.8	17.4	13.7	19.7	17.0	18.6	16.4	17.9	13.7	**389.2**
NSF	150	204	221	256	335	359.7	388.8	408.6	408.6	428.7	485.1	466.3	421.0	410.6	412.4	**5355.8**
USDA (total)	0	0	0	2	3	6.2	6.8	10.1	15.3	20.313	20	18.317	19.498	19.07	18.8	**159.4**
ARS	0	0	0	0	0	0	0	0	0	0	0	2.0	2.0	2.0	2.0	8.0
FS	0	0	0	0	0	2.3	2.9	4.6	5.4	7.1	10.0	5.0	5.0	4.0	4.0	50.3
NIFA	0	0	0	2	3	3.9	3.9	5.5	9.9	13.2	10.0	11.3	12.5	13.1	12.8	101.1
TOTAL†	**464**	**697**	**760**	**989**	**1200**	**1351.3**	**1424.1**	**1554.5**	**1701.5**	**1912.8**	**1847.3**	**1857.3**	**1550.2**	**1537.5**	**1536.9**	**20383.4**

* NOTE: Totals may not add due to rounding

† NOT including American Recovery and Reinvestment Act funds for NSF ($101 M), DOE ($293 M), NIST ($43 M), and NIH ($73 M)

†† FY '14 estimated based on 2014 enacted levels and may shift as operating plans are finalized

††† FY '15 Request

Table 2. FY2015 request for funding by Program Component Area from the National Nanotechnology Initiative Supplement to the President's 2015 Budget

Proposed 2015 Agency Investments by Program Component Area (dollars in millions)											
	1. Nanotechnology Signature Initiatives (NSIs)	1a. Solar Energy	1b. Nanomanufacturing	1c. Nanoelectronics	1d. NKI	1e. Sensors	2. Foundational Research	3. Nanotechnology-Enabled Applications, Devices, and Systems	4. Research Infrastructure and Instrumentation	5. Environment, Health, and Safety	NNI Total
CPSC	**0.0**	0.0	0.0	0.0	0.0	0.0	**0.0**	**0.0**	**0.0**	**2.0**	**2.0**
DHS	**0.0**	0.0	0.0	0.0	0.0	0.0	**0.0**	**24.4**	**0.0**	**8.0**	**32.4**
DOC/NIST	**24.4**	3.1	6.1	11.5	1.3	2.4	**11.3**	**4.6**	**35.6**	**6.7**	**82.6**
DOD	**47.3**	2.3	0.8	22.0	0.9	21.3	**72.2**	**22.4**	**0.0**	**2.1**	**144.0**
DOE	**42.8**	35.3	0.0	0.0	5.0	2.5	**136.7**	**29.2**	**134.4**	**0.0**	**343.1**
DOT/FHWA	**0.0**	0.0	0.0	0.0	0.0	0.0	**0.0**	**1.5**	**0.0**	**0.0**	**1.5**
EPA	**0.0**	0.0	0.0	0.0	0.0	0.0	**0.0**	**0.0**	**0.0**	**16.8**	**16.8**
DHHS (total)	**49.1**	0.0	0.0	0.0	0.0	49.1	**119.0**	**228.1**	**19.9**	**53.4**	**469.6**
FDA	0.0	0.0	0.0	0.0	0.0	0.0	0.0	0.0	0.0	17.0	**17.0**
NIH	49.1	0.0	0.0	0.0	0.0	49.1	119.0	228.1	19.9	25.3	**441.5**
NIOSH	0.0	0.0	0.0	0.0	0.0	0.0	0.0	0.0	0.0	11.1	**11.1**
NASA	**6.4**	0.1	3.9	0.0	0.0	2.4	**5.6**	**1.3**	**0.4**	**0.0**	**13.7**
NSF	**115.6**	27.7	23.4	38.0	19.0	7.5	**181.4**	**46.6**	**46.5**	**22.3**	**412.4**
USDA (total)	**5.7**	0.7	2.0	0.0	0.0	3.0	**4.2**	**4.9**	**3.0**	**1.0**	**18.8**
ARS	0.0	0.0	0.0	0.0	0.0	0.0	0.0	2.0	0.0	0.0	**2.0**
FS	1.0	0.0	1.0	0.0	0.0	0.0	2.0	0.0	1.0	0.0	**4.0**
NIFA	4.7	0.7	1.0	0.0	0.0	3.0	2.2	2.9	2.0	1.0	**12.8**
TOTAL	**291.3**	69.2	36.2	71.5	26.2	88.2	**530.4**	**363.0**	**239.8**	**112.4**	**1536.9**

The 2014 NNI review report is organized into Sections around categories of recommendations. The Sections following this introduction are organized as follows:

1) NNI Goals and Progress–the status of the science
2) Grand Challenges for NNI 2.0 – the mechanism behind NNI 2.0
3) Program Management–the organizational structure of the Federal component of NNI 2.0
4) Commercialization – the path to commercialization needed under NNI 2.0
5) Research Enterprise–the discovery and exploration breakthroughs needed under NNI 2.0
6) Environmental, Health, and Safety Issues – the development of an EHS strategy under NNI 2.0

We also include an appendix and offer an example of how to measure the continuum of the research / translation process and subsequent outcomes.

Each of the Sections relate back to the core premise of this report, that a Grand-Challenge framework will provide the necessary mechanism to motivate stronger cross-agency partnering and the building of an ecosystem bringing together government, industry, and academia. In the end, if the lines between nanotechnology and other technologies are blurred in the drive to fulfill these Grand Challenges and nanotechnology becomes part of the greater research and translation portfolio, NNI 2.0 should be viewed as a great success.

SECTION 2. NNI GOALS AND PROGRESS

From the outset, the NNI was envisioned with broad and far-reaching ambitions, with the ultimate outcome of societal benefits[10]. Four goals were defined:

1) To advance world-class nanotechnology research and development;
2) To foster the transfer of new technologies into products for commercial and public benefit;
3) To develop and sustain educational resources, a skilled workforce, and the supporting infrastructure and tools to advance nanotechnology; and
4) To support the responsible development of nanotechnology.

Over twelve years have passed since these NNI ambitions were established, and it is fair to assess how well the United States has progressed towards these aims. While perfect metrics to report progress towards these goals do not exist, this review reiterates the need to develop measurements that can be applied to specific Federal program elements while drawing on available measurements to judge overall NNI progress[18]. A brief assessment is performed here with the goal of calibrating the magnitude of achievement.

With respect to Goal 1 (to advance world-class nanotechnology research and development), Figure 3 reports on nanotechnology research and development through the number of papers and patents published by authors in the United States and other leading nanotechnology regions[19]. The dramatic increase in published papers demonstrates the United States' commitment to nanoscience research over the life of the NNI. While the rate of patents published in three or more countries has decreased over the past decade, the cumulative growth of nanoscience patents follows an S-curve technology-development model suggesting that the field of nanotechnology product development is growing toward, but has not yet reached, maturity[20]. Together, these data demonstrate the clear creation of nanoscience knowledge and the United States' position as a global leader in the field.

With respect to Goal 2 (to foster the transfer of new technologies into products for commercial and public benefit), Figure 4 reports on the commercialization of nanotechnologies using the measured (2003-2009) and predicted (2010-2015) nano-enabled products revenue in the United States and other global regions[22]. Although some market revenue has been demonstrated, these data suggest that the full market potential of nanoscience has yet to be achieved. The magnitude of the revenue predicted – up to $20 billion in the United States in 2015—isconsiderable, demonstrating that nanotechnology has the potential to contribute significantly to our economy. Again, the United States is a leader in the field, with Europe and the Asia-Pacific region as close competitors.

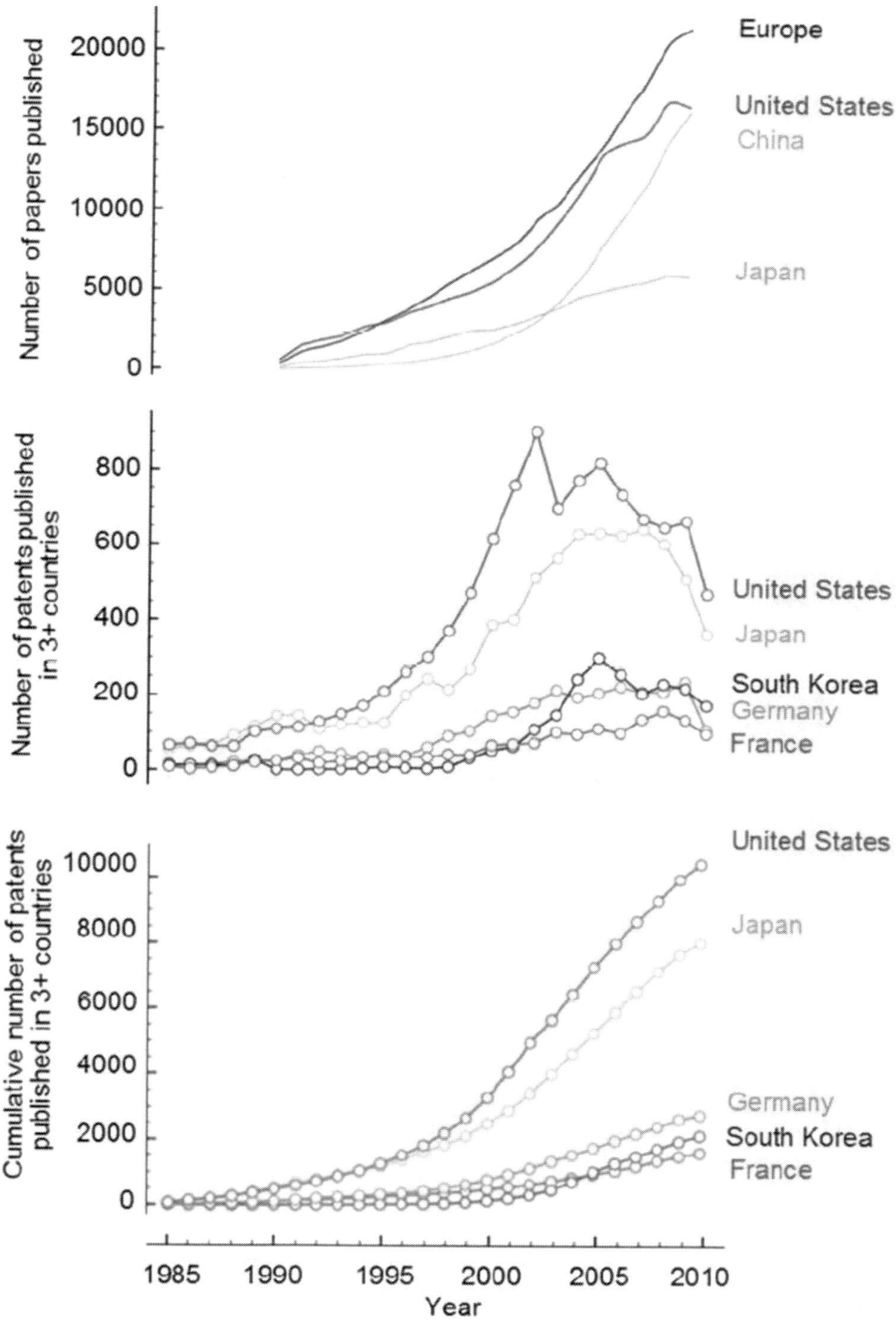

Figure 3. NNI Goal 1. Number of published papers and patents for the United States and other international leaders. The metric of patents published in three or more countries is a more representative indicator of signißcance compared with patents published in just one or two countries[19,21].

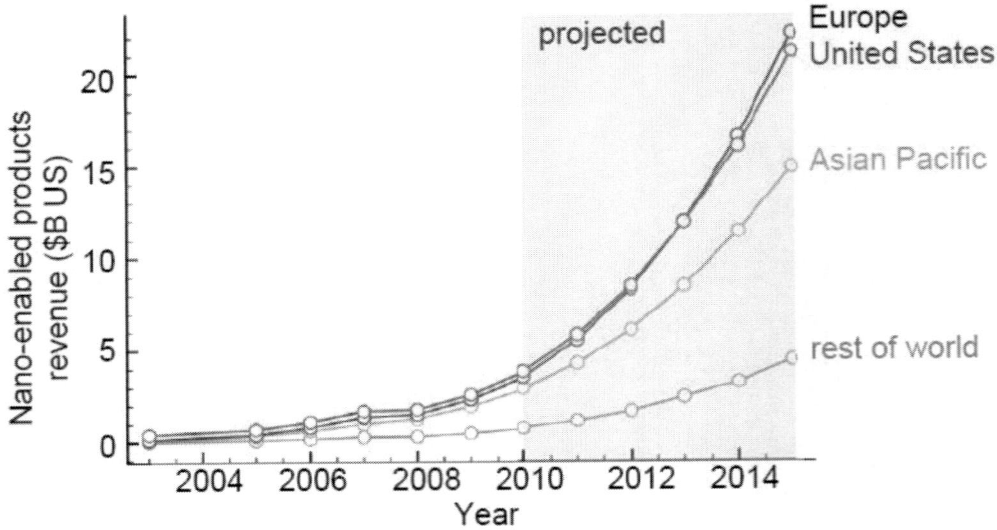

Figure 4. NNI Goal 2. Nano-enabled products in the United States and other regions are increasing their market revenue, with signißcantly increasing growth rates predicted for the coming years. Data from Lux Research.

With respect to Goal 3 (to develop and sustain educational resources, a skilled workforce, and the supporting infrastructure and tools to advance nanotechnology), Figure 5 reports on the development of the nanotechnology workforce and infrastructure. The number of existing Federally funded nanoscience centers, typically large-scale endeavors where faculty and students from many departments and even different institutions share ideas and collaborate, can be used to indicate educational and workforce development activities. The number of centers increases over the years and starts to plateau in 2010. The reported data take into account both the creation of new centers, as well as the sunsetting of centers for which funding has expired[23]. The number of users at Federally funded nanoscience-research facilities, which often house advanced equipment for nanoscale research – for example, transmission and scanning electron microscopes, various spectroscopic instruments, and electron and ion beams for lithography– can be used to indicate the creation of infrastructure and tools, as well as education, training, and workforce development. The Department of Energy Nanoscale Science Research Centers (DOE NSRCs) have increasingly provided equipment, training, and access to users. DOE reports up to roughly 400 users per center in 2012, an average of one user per day for a given center.

Figure 6 attempts to assess Goal 4 regarding the responsible development of nanotechnology. Federal investments in nanoscience research related to Environmental, Health, and Safety (EHS) and Educational, Ethical, Legal, and Societal Issues (EdELSI) are plotted in both absolute dollars and as a percent of total NNI funding. These Federal investments certainly demonstrate "support." The responsible development of nanotechnology has not been fully achieved – consider, for example, the lack of clear regulations and the many unknowns regarding EHS effects emerging at the nanoscale. While some progress has been made towards the five research categories identified in 2008 as NNI EHS priority strategies, the sheer range of chemical compositions, sizes, and shapes encompassed in the category "nanomaterials" ensures a considerable task in accurately characterizing each material and ultimately developing effective EHS policies[24]. Much work remains to be done in this area.

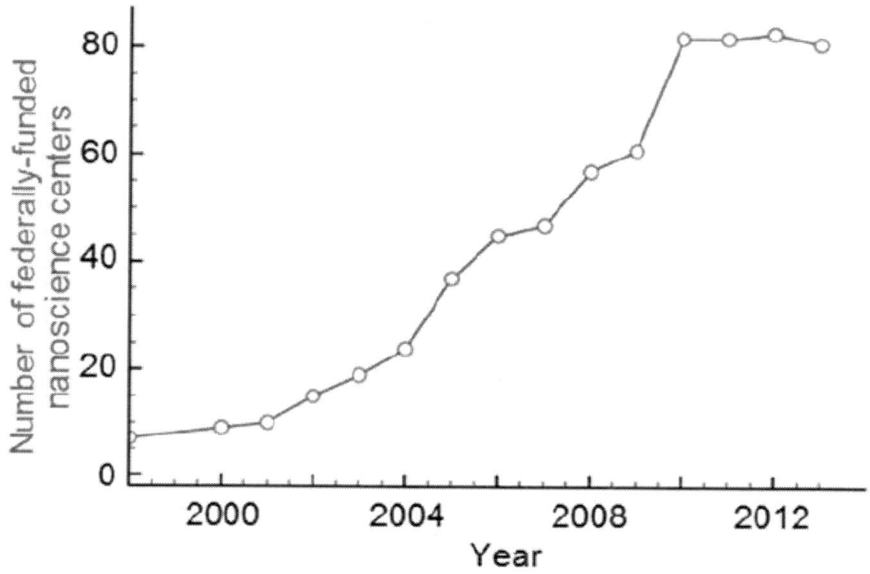

Figure 5. NNI Goal 3. Infrastructure and workforce training has developed in the United States over 12 years, with the growth of Federally funded nanoscience centers and user facilities. Data gathered from NNI and Federal agency websites.

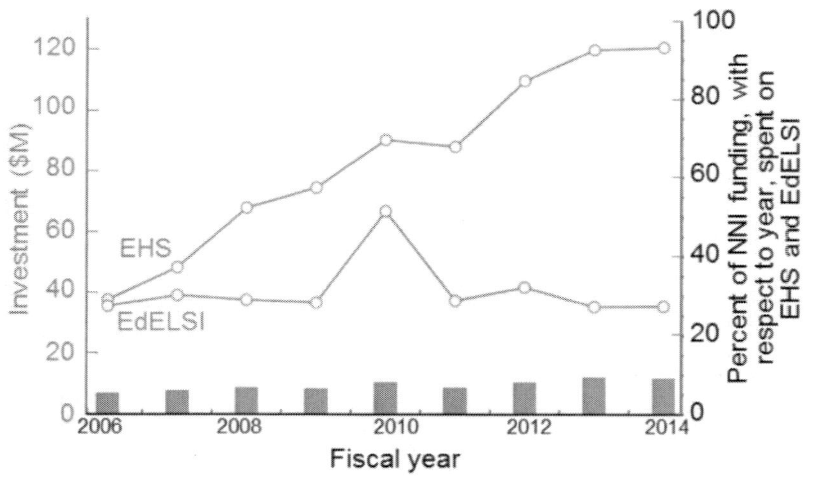

Figure 6. NNI Goal 4. U.S. EHS and EdELSI investments in dollars (left ordinate and traces) and as a percentage of total NNI funding (right ordinate and bars). FY2009 and FY2010 data include ARRA funding; FY2013 and FY2014 data report estimated and requested amounts, respectively. Data gathered from http://nanodashboard.nano.gov.

The data presented here provide one view of NNI accomplishments, and they demonstrate that clear progress has been made. But much work remains in order to accomplish the goals of the NNI, which was conceived as a long-term endeavor[25]. The development of metrics that more fully capture NNI achievements, coupled with the collection of data for these metrics, will provide an increasingly accurate assessment of the NNI.

International Collaboration and Competition

The Federal Government's nanotechnology research, commercialization, regulation, and ultimately societal benefit unfold in a global context. A 2014 evaluation of nanotechnology publications for 2011-2013 shows that China and the European Union have led the United States in numbers of nanotechnology publications, with China publishing the most papers in nanotechnology for the first time in 2013 (Figure 7). Increases in spending overseas have made nanotechnology commercialization and particularly nanotechnology manufacturing a national priority in many countries. Over 60 countries have invested in nanotechnology at the national level[19]. As of 2012, the top four individual countries with national nanotechnology investments were the United States (Federal and states $2.1 billion), Japan ($1.3 billion), Russia ($974 million), and Germany ($617 million). Creating and maintaining a separate, coordinated nanotechnology effort at a national scale, as in the United States and Germany, is an approach some countries, such as Japan and Russia, have moved away from as initiatives expire and nanotechnology applications become integral to multiple facets of broader science and technology funding. In 2011, Japan cut nanotechnology as a focus area and integrated it with the larger funding plan while continuing to increase federal nanotechnology support from $918 million in 2010 to $1.3 billion in 2012.

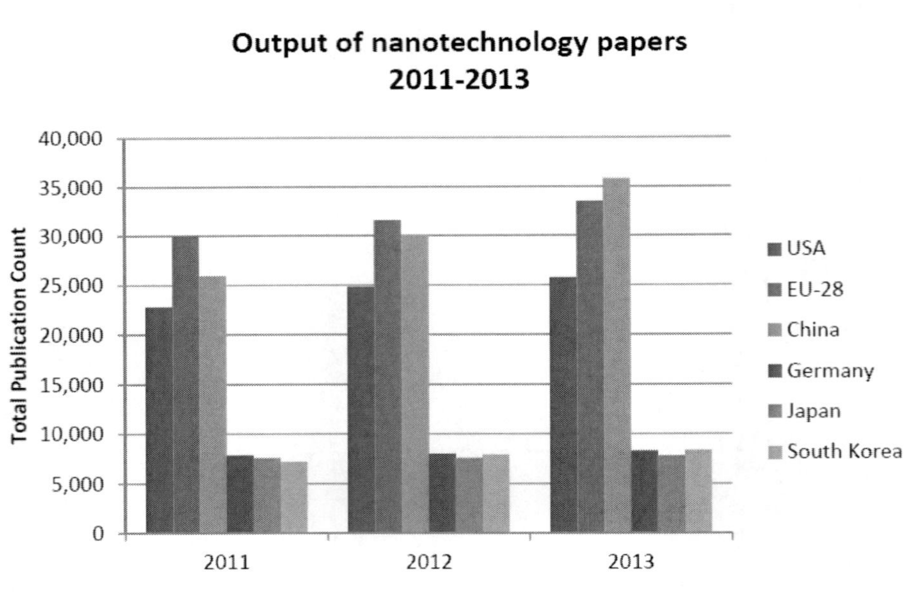

Figure 7. From 2011 to 2013, the United States, European Union, and China produced the largest numbers of nanotechnology papers published each year, with China leading[26,27].

Another important direction of NNI 2.0 leadership will be to help determine where the United States best competes with other countries and where it should collaborate. The NNI Strategic Plan released by the Office of Science and Technology Policy (OSTP) in February 2014 remarks that the United States should be actively engaged "in international activities integral to the development and responsible commercialization of nanotechnology-enabled products and processes."[28] This includes (i) participating and leading the development of international nanotechnology standards; (ii) engaging in bilateral and multilateral collaborations and cooperative activities to further nanotechnology-related commercialization, innovation, and trade; and (iii) supporting forums in which U.S. and international stakeholders can exchange technical information and discuss market needs, intellectual property rights, and other issues relevant to enabling commercialization.

The call for collaboration supports the development of revolutionary new technology that has far-reaching implications for innovation, economic development, and the advance of sustainable scientific enterprises. Accordingly, OSTP has bilateral science agreements with Russia, China, India, Brazil, Japan, and South Korea. The Unites States' international participation in nanotechnology activities benefits all sides in knowledge exchange, increased trade, and economic growth, and it also catalyzes U.S. leadership in a growing global nanotechnology enterprise that is expected to attain a global value of $4.4 trillion by 2018[22].

The DOE Nanoscale Science and Engineering Centers, for example, accept proposals to conduct fundamental research in the Centers regardless of the proposal's country of origin. The winning proposals are those with the highest scientific merit, bringing the highest quality research to the United States.

The United States also benefits from international collaboration by attracting the brightest and best-educated graduates from outside the country to work in U.S. universities and industry. In the United States, they infuse their advanced knowledge, skills, and fresh ideas into the U.S nanotechnology enterprise. Because of the breadth and depth of the U.S. research enterprise and translation of intellectual property into new commercial opportunities, international collaboration has contributed to expansion of the global nanotechnology enterprise and trade opportunities that has benefited the U.S. economy.

While the development of electronics, energy generation, batteries, supramolecular chemistry, synthesis through self-assembly, biotechnology, and nanomedicine are transparent worldwide academic and industrial enterprises, the advent of molecular nanofabrication could drastically change the balance of leadership in nanotechnology. It will require extraordinarily complicated and revolutionary advances in knowledge, but the ability to produce nano-inspired molecular manufacturing could spawn a technical revolution. As with all dual-use technologies, the impact of malicious use must be considered.

How should the United States deal with these contrasting scenarios in evaluating its international collaborations and agreements? Staying competitive and actively engaged in the international nanotechnology enterprise would appear to be the best option, not only from a global economy perspective, but also having access to the ever-growing nanoscience and nanotechnology knowledge base and toolbox. This will allow the United States to develop its own competitive nanomanufacturing enterprise and accompanying awareness of the development of asymmetrical capabilities that may require further scrutiny.

Next Generation Nanotechnology Research

Balancing the focus on Grand Challenges and with the awareness of collaborative and competitive international engagement, NNI 2.0 will need to continue to invest in and coordinate discovery and exploratory research in nanotechnology. Research and development in nanotechnology during this second phase of investment will seek to bring reproducible atomic-scale control to nanoscale building blocks, build active nanostructures instead of passive ones, and integrate nanoscale elements into complex assemblies with emergent and, ultimately, predictable properties. Nanotechnologies with these properties could transform energy, advanced materials, electronics, medicine, and more. The technical challenges in achieving control, function, and integration of nanoscale building blocks, however, will likely be more difficult than the creation of the nanoscale units themselves. Drawing an analogy to electronics, the discovery of the transistor was only the first step in a long journey to laptops, smartphones, and the internet. The development of a stable workforce and infrastructure is a responsibility that NNI 2.0 should prioritize to be fulfilled by the agencies and the private sector. Many of competitors to the United States have established long-term strategic goals that range up to 2025.

SECTION 3. GRAND CHALLENGES FOR NNI 2.0

Additional leadership and management tools are needed to bring about the transition to NNI 2.0 advocated by PCAST. We recommend that the construct of Grand Challenges be instantiated across the NNI ecosystem and in the management of Federal activities to focus NNI participants on significant problems of major national interest that, by commercializing the associated science and technology, will benefit society. Organizing activities around Grand Challenges can be a major community rallying point and provide additional tools to manage and measure the effectiveness of NNI 2.0.

This Section provides the details behind this proposed Grand Challenges-based model. To ground the discussion, we first look at one current programmatic element of NSET's activities, the Nanotechnology Signature Initiatives (NSIs). We review progress and effectiveness of the implementation of NSIs and then describe in detail the rationale for and characteristics of Grand Challenges, how they differ from the NSIs, and illustrative examples that could be applicable for NNI 2.0 Grand Challenges. Specific recommendations are then made to instantiate the Grand Challenges construct, with particular focus paid to how to engage the NNI community in identifying and selecting Grand Challenges.

Nanotechnology Signature Initiatives

Beginning with its 2010 strategic planning process, the NSET created focal areas for cross-agency collaboration and coordination, culminating in the creation of five Nanotechnology Signature Initiatives[29]. The NSIs define the shared vision of OSTP and participating Federal agencies for advancing nanoscale science and technology from research to commercialization through enhanced interagency coordination and focused investment.

NSIs are selected as broad areas that exhibit particular promise and significant opportunity and that involve three or more Federal agencies. Five topics have been chosen as detailed in the 2014 NNI Strategic Plan[28]. Each is described in a white paper that includes four elements: national need, "thrust areas" or goals, expected outcomes, and agency roles and contributions[30]. The five current NSIs are:

- Nanotechnology for Solar Energy Collection and Conversion: Contributing to Energy Solutions for the Future,
- Sustainable Nanomanufacturing: Creating the Industries of the Future,
- Nanoelectronics for 2020 and Beyond,
- Nanotechnology Knowledge Infrastructure: Enabling National Leadership in Sustainable Design, and
- Nanotechnology for Sensors and Sensors for Nanotechnology: Improving and Protecting Health, Safety, and the Environment.

Within its mandate to review the NNI triennially, the National Research Council Committee in 2013 assessed the framework, planning, and management of the NSIs. The Committee specifically examined the practices, agency collaborations, and progress of the 2011 NSIs and analyzed all five NSI white papers for the inclusion of six characteristics of an effective strategy identified by the Government Accountability Organization (GAO). Table 3 shows the historical, current and requested funding for the NSIs. They determined that only the first two GAO characteristics – purpose, scope, and methods; and problem definition and risk assessment – are addressed adequately by the white papers[31].

Table 3. Total funding for Nanotechnology Signature Initiatives 2011-2015 (dollars, rounded to the nearest million)[16,28]

	2011 ACTUAL	2012 ACTUAL	2013 ACTUAL	2014 EST.	2015 PROPOSED
Nanomanufacturing	61	56	35	38	36
Solar Energy	88	88	74	67	69
Nanoelectronics	97	92	87	77	72
Nano Knowledge Infrastructure	-	2	8	32	26
Sensors	-	55	77	84	88

The Committee observed that the NSIs are "obvious vehicles for collaboration with the private sector" and highlighted the industry-led Nanoelectronics Research Initiative as a model partnership. They also recommended that each NSI team implement formal interagency plans and expand the white papers to include a strategic plan incorporating the elements described by the Committee. Further, the Committee put forward the success of the NSIs as a measure of success for the NNI and suggested that the NSIs would benefit from roadmapping activities, efforts to tie other NNI activity to NSI goals, and improved communication with researchers whose projects contribute to the NSIs.

Characteristics of a Grand Challenge

While an NSI is shaped by the coordination of three or more Federal agencies around a topic, a Grand Challenge is a large, outward-facing effort with a specific, measurable goal. A Grand Challenge has a well-defined technical goal with a story-telling case that inspires different sectors to invest in achieving the goal. Most Grand Challenges address an issue of significant societal impact. An example is the goal of the DOE SunShot Initiative: "reduce the total costs of photovoltaic solar energy systems by about 75% so that they are cost competitive at large scale with other forms of energy without subsidies before the end of the decade."[32] A nanotechnology Grand Challenge should be audacious but achievable and stimulate the network of activities that will drive scientific ideas to commercial nanotechnology and catalyze new discovery for technologies of the future.

Grand Challenges exhibit other important characteristics:

- They have a measurable end-point. It is clear when they have been reached. As such, they also have a finite, albeit relatively long (probably a decade), lifetime.
- They require advances in fundamental scientific knowledge, tools, and infrastructure for successful completion. In short, when a Grand Challenge is begun, all the resources needed to complete it are not known. As such, it is necessary to recognize and articulate the risks of the undertaking and to mitigate those risks to the maximum extent possible.
- There must be clear milestones en route to the final Grand Challenge goal that are both measurable and valuable in their own right. It is only through monitoring these deliverables that it is possible to tell whether or not the effort is on track to achieve its ultimate objective.
- They are integrating. Their solutions require bringing together multiple disciplines – in many cases, disciplines that do not typically interact. In addition, Grand Challenges span from fundamental science to engineering demonstration and, upon completion, to commercialization.
- They are too big to be undertaken by a single, or even a few, institutions. In fact, one way of mitigating the risk inherent in taking on an effort of this magnitude may be to pursue more than a single approach to the problem, thus involving even more institutions than would be engaged in a single approach.

Recommendation 1. Establishing Grand Challenges is an effective means for focusing and amplifying the impact of Federal nanotechnology activities. NSET and OSTP should identify a list of candidate nanotechnology Grand Challenges that address significant societal needs. At least one Grand Challenge should contain program elements aimed at manufacturing challenges specific to that focus area.

Identifying and Implementing Grand Challenges

NSET, the NSTC Committee on Technology, and OSTP have key roles in creating and implementing the process to identify Grand Challenges. This process needs to provide context for the conversations and deliberations that lead up to Grand Challenge selection. Since Grand Challenges will have different lifetimes, depending on their scale and complexity, as well as the resources available to each, Grand Challenge selection and review will be a process that occurs periodically, probably annually.

> **Recommendation 2.** NSET and OSTP should create and execute a process to engage research, development, and industrial stakeholders in the identification and selection of Grand Challenges on an ongoing basis.

Essential elements to identify and implement Grand Challenges include:

- The investment of the public and industrial, academic, national laboratory, investor, financial, and communication sectors.
- A strong leader who is a member of NSET and who can set a vision for a Challenge and convene stakeholders toward its development.
- Identification of critical challenges in the mission space of agencies participating in NNI that have a solution requiring significant advances in nanoscience and technology.
- Understanding of the global landscape in the problem area. What countries are working in the area? How far along are they? What are the prospects of their solving the problem first? If they are ahead, is it likely that we could partner with them, should that be desirable? If they are ahead, is it of critical national importance that we catch up with and/or surpass them?
- Engage broad swaths of stakeholders in the dialogue leading up to Grand Challenge selection. This includes researchers, research managers/leaders, and agency representatives.
- After allowing for significant community engagement, a fairly small set of subject-matter experts and senior advisors should select the Grand Challenges.

We strongly support the participation of the agencies in developing Grand Challenges, including a regular, structured process for creating and supporting new Grand Challenges on an annual basis and reviewing the progress on existing Challenges. One component should be brainstorming workshops where industry representatives, top junior and senior nanoscientists and nanotechnologists, venture-capital firms, and Federal agencies get together with other relevant groups to (a) define a Grand Challenge, (b) identify the roadblocks to progress, and (c) propose ways to address these challenges.

The Federal Government has several possible strategies that could help the nanotechnology community reach the Challenges. The Nanotechnology Faculty Fellows program discussed in Section 6 on Research Enterprise could provide a natural leadership base for the conduct of such meetings and evaluation of the ideas developed. Federal agencies can offer innovation prizes that reward the first person or group that achieves one milestone of a Grand Challenge, and they can offer matching funds to competitive applications for public-private partnerships.

Recommendation 3. Federal agencies, with support from OSTP, should offer implementation tools like innovation prizes and public-private partnerships to encourage researchers to reach critical milestones on the path to completing Grand Challenges.

Illustrative Grand Challenges

Thanks to the foundation laid by the NNI, the United States is in a position to move to a new level of integration in nanoscale science and technology in which societal challenges requiring nanoscale science and technology solutions can be addressed by bringing to bear a disciplined, coordinated effort spanning from fundamental scientific discovery to engineering demonstration and commercialization. While the specific challenges to be addressed in the next generation of NNI should be articulated by a process to be developed and implemented by agencies, NSET, and OSTP, a few examples of possible Grand Challenges may help to illustrate the points of this Section. ***The example Grand Challenges in Boxes 3-6 are only illustrative: actual Grand Challenges will need to be much more clearly articulated and focused, will need to be determined only after extensive community engagement, and will need have their scope and resources carefully aligned***.

Box 3. Illustrative Nanotechnology Grand Challenge #1

Nano-enabled Desalination of Seawater to Solve the Emerging Water Crisis.

The Organisation for Economic Co-operation and Development and the United Nations have reported that 350 million people in 25 different countries are currently suffering from water shortage, and this will grow to 4 billion people (two-thirds of the world population) in 52 countries by 2025. The lack of clean water exacerbates health and energy challenges, making the need to produce clean water one of the most pressing challenges of the 21st century. In most of the world, desalination of sea water is the only possible solution, but the process of desalination remains too energy-intensive to be practical. Different approaches have been followed in trying to use nanotechnology for desalination, including the incorporation of nanoparticles into existing conventional thin-film composite reverse osmosis (RO) membranes. An alternative approach is to use nanotechnology to create novel nanostructures such as graphene and carbon nanotubes, which could be used as a replacement RO membrane with a highly organized ultrathin structure. These new membranes offer very high selectivity in separating out the salt as the pore morphology can be precisely controlled and energy use can be minimized.

The grand challenge is the reduction in the specific energy consumption of seawater desalination to below 1.5 kWh/m^3.

Box 4. Illustrative Nanotechnology Grand Challenge #2

Reducing Greenhouse Emissions with Nano-enabled Solid-State Refrigeration.

Primary power generation to produce the energy required to run heating, ventilating, and air conditioning (HVAC) systems in the commercial and residential built environment is responsible for a significant portion of U.S. greenhouse gas emissions.

Solid-state refrigeration holds the promise of curtailing greenhouse gases, reducing system level noise, and optimizing unit size, weight, and power by more efficiently cooling commercial systems. Thermoelectrics, thermoelastics, electrocalorics, thermionic, and magnetocalorics are the solid-state materials that underlie these systems. The engineering challenge is to improve the materials' performance by reducing the energy lost when energy is converted for cooling in the material (caloric cooling) and by decreasing energy lost at the interfaces between the material and its device and between a single device and the whole system (parasitic loss). Those losses are defined in a Coefficient of Performance (COP). Tailoring materials at the nanoscale could be the basis of significantly superior structures that have higher caloric conversion and reduced parasitic losses.

The Grand Challenge is to develop solid-state cooling systems with COP > 5 (25% higher than current HVAC technologies) and power density > 20 kW/ m^3 (two times greater than current HVAC technology). This will lead to commercial systems that have the potential to reduce greenhouse gas emissions from climate control in buildings by as much as a 20%, resulting in an annual reduction of nearly 200 million metric tons of carbon equivalent emissions, equivalent to removing approximately 40 million passenger cars from U.S. highways.

Box 5. Illustrative Nanotechnology Grand Challenge #3

Creating a New Forefront of Manufacturing through Nano 3D Printing.

Three-dimensional (3D) printing is at the forefront of the manufacturing revolution and has the potential to transform U.S. manufacturing and prototyping capabilities in fields as diverse as personalized medical care and vehicle design. The resolution, throughput, and materials flexibility of 3D printing, however, is limited at present. Over the next 10 years, the scientific community should develop 3D-printing technology with the materials flexibility of chemical synthesis and the resolution and throughput of high-cost semiconductor nanofabrication tools in order to create on-demand goods with unique and desirable functions.

Fabrication technologies such as injection molding have the ability to mass-manufacture goods with resolution in the range of hundreds of microns down to hundreds of nanometers; this technique is often cost prohibitive, however, for rapid prototyping and individualized goods. 3D printing has shown great potential in overcoming these limitations, and within the next decade, 3D printing could manufacture large-scale prototypes with precise control over chemical composition, with resolution matching that of modern fabrication techniques, and at a speed and cost that would facilitate wide-spread adoption and household use. With these design aspects perfected, 3D printing will be able to surpass traditional fabrication techniques in the fidelity of manufacturing to design and speed up the manufacturing process. Potential benchmarks toward this goal may be the production of personalized medical devices (e.g., heart stents, replacement joints) at competitive prices to current techniques.

Box 6. Illustrative Nanotechnology Grand Challenge #4

Development of a Nanoscale Therapeutic for at Least One Major Cancer Type by 2030

In spite of impressive advances in our understanding of fundamental cancer biology and the development of a new generation of targeted cancer therapies, the achievement of long-term survival of most common cancers remains elusive. Multiple factors limit the effectiveness of cancer treatment strategies, including: late disease detection, complex and heterogeneous tumor physiology that necessitates combination regimens, metastasis, natural tissue barriers and solid tumor structures that impede drug penetration, and toxicities that result from exposure of healthy tissues to these therapies. Nanotechnology has the potential to significantly improve upon current cancer therapies through the use of "smart" nanoparticles that specifically address these hurdles.

The focus of this Grand Challenge is on the development of novel, multifunctional nano-carriers capable of efficiently transporting diagnostic probes and therapeutic payloads across tissue barriers to specific tumor types. This next generation of carri- ers will require flexible payload capabilities and highly effective, tumor-specific tar- geting strategies. The successful development of such a technology platform would enable significant improvements in early cancer detection through tumor-selective delivery of diagnostic probes and imaging agents. The vehicle would also be used to deliver optimized drug combinations directly to the tumor, thereby enhancing thera- peutic efficacy and minimizing exposure of healthy tissues. A 10-year horizon is anticipated for this work. A key early milestone is the selection of cancer type on which to focus. Pending the development of robust tumor target- ing strategies, nano-based sensors that significantly increase biomarker detection sensitivity and nanoscale tumor imaging will improve early diagnosis within 5 years. This will enable the development of multifunctional, engineered nanocarriers to deliver drug combinations (drugs, nucleic acids, etc) directly to the tumor. The goal is to deploy a 10-fold more sensitive early detection technology for a major can- cer type within 5 years and clinically validate a next-generation nanotechnology med- icine delivery system that achieves significant improvement in long term survival in the next 10 years.

Section 4. Program Management

Focusing on Grand Challenges

The decentralized nature of the Federal Government's activities in the National Nanotechnology Initiative has motivated PCAST and others to express concern in several reviews, particularly related to setting priorities and coordinating around major themes and opportunities. The enhanced focus on commercialization we argue for in this report reinforces those concerns. Each Federal agency investing in nanotechnology determines its own budget, which is either allocated specifically for nanotechnology or blended within its larger science and technology portfolio. The steering and coordinating bodies of the overarching Federal process – the Nanoscale Science, Engineering, and Technology Subcommittee (NSET) and the National Nanotechnology Coordination Office (NNCO), respectively– are not empowered to make decisions about resource allocations across or within agencies, and the Committee on Technology (COT) in the National Science and Technology Council does not play a leadership role setting common interagency goals and ensuring they are met (Figure 2). NSET convenes agency representatives, but it does not manage agency activities in an integrated way nor create a unified plan within which a specific agency can manage its activities. This loose coupling may have sufficed when the emphasis was on broadening and deepening nanotechnology research, but without a more coordinated approach to commercial transition, the promise of nanotechnology is likely to be left unrealized. The nanotechnology investment by the United States comes at a high opportunity cost when research occurs but no path is available for commercialization. It is within this landscape that PCAST, as described in the previous Section, recommends the deployment of Grand Challenges. In addition to the energy it will create in the broad nanotechnology community, we believe the Grand Challenge framework holds substantial potential to enable more explicit and, we believe, more effective leadership and management of NNI 2.0. The specific, common goal of a particular Grand Challenge requires a more active management. By publicly laying out detailed technical and programmatic goals and creating a process in which stakeholders debate and invest in them, defining a Grand Challenge creates a mechanism for an agency, or multiple agencies working in a coordinated, funded fashion, to set priorities. It also helps to ensure that the Federal investment in nanotechnology associated with the Grand Challenge goal becomes a major element of a larger public and private investment and a more productive nanotechnology ecosystem that can support the efficient commercialization of nano-based products. NNI 2.0 leadership thus will not only ensure the introduction of novel nano-based materials, medicines, and technologies but will enhance U.S. competitiveness in a key technology area and lead to infrastructure, capability, and workforce improvements, critical areas of national interest.

Program Management

This section of the report looks at the state of the tools, methods, and responsibilities employed to manage the NNI. Recommendations are made to provide the more active leadership we believe is necessary to successfully transition to NNI 2.0. These recommendations are both enabled by the Grand Challenges process and critical to its success. This theme of more active leadership is not new, and we begin by providing in Table 4 an assessment of the responses to recommendations regarding management and leadership that were made in the 2012 version of this report.

Table 4. Response to 2012 PCAST recommendations on program management

2012 NNI Review recommendation	Activities in response
The NNCO in partnership with the Office of Science and Technology Policy (OSTP) should work with the agencies to develop agency implementation plans for achieving the goals and objectives outlined in the 2011 NNI strategic plan.	The 2014 NNI Strategic Plan updates and replaces the 2011 Plan and continues "the integrated, interagency approach that informs the nanotechnology-specific strategic plans of the agencies."

Table 4. (Continued)

2012 NNI Review recommendation	Activities in response
	There is no evidence that the NNCO and OSTP worked directly with NNI agencies to develop formal, agency-specific strategic plans to achieve the NNI objectives, reflecting a preference to defer to agencies for such activity.
Participating agencies should ensure that senior agency officials capable of influencing funding deci-sions are participating fully and personally in stra-tegic planning activities of the NSET. Officials at this level, in contrast with representatives active at the program or office level, could more effectively drive agency planning and budget allocations to meet NNI strategic directions.	The revised NSET charter stipulates that each agency will appoint one principal representa-tive with the authority to represent the agency. The NSET membership lists in the NNI Budget Supplements, however, show minimal change in NSET representation after the 2012 PCAST recommendation, and the involvement of sen-ior agency officials on the membership list in NSET activities cannot be determined. The re-vised charter also describes a Senior Steering Group (SSG) that will meet "as needed and at the discretion of the NSET co-chairs." There is no publicly available information to confirm that the SSG has been established or has met.
The Nanotechnology Signature Initiatives should be fully supported in NNI budgets. To this end, PCAST recommends that the Office of Management and Budget increase funding to these Initiatives.	Investment in most Nanotechnology Signature Initiatives, as reported in the annual NNI Supplement to the President's Budget, is essentially flat with the exception of the continued decline in nanoelectronics and robust increase in sensor technology investment (see Table 3). In-creased funding had been requested; flat fund-ing in a tight budget environment is consistent with the spirit of the recommendation.
The NSET Subcommittee should create Nanotechnology Signature Initiatives in other priority areas such as homeland security, national defense, and human health.	Launched in 2012, the NSI Nanotechnology for Sensors and Sensors for Nanotechnology: Improving and Protecting Health, Safety, and the Environment is led by the NIH, which reports a $49 million investment. The chemical and biological sensor technologies developed through this NSI have significant applicability to home-land security in addition to health and the envi-ronment. The NSI Nanotechnology Knowledge Infrastructure: Enabling National Leadership in Sustainable Design, also launched in 2012, uti-lizes human-health data in its initial efforts. DOD participates in mission-related aspects of all existing Signature Initiatives.
Appoint the NNCO director as co-chair of the NSET Subcommittee of the NSTC.	The role of the NNCO director is unchanged in the revised charter. The Director continues to manage the NNCO as the secretariat to the NSET Subcommittee and participates in NSET Subcommittee meetings as a committee mem-ber but without the policy role implied by NSET co-chairmanship.

2012 NNI Review recommendation	Activities in response
Change the requirement that the NNCO director must come from within the Federal Government to allow external, non-Federal experts the opportunity to direct the NNCO.	The revised NSET charter includes the use of the Intergovernmental Personnel Act Mobility Program to staff the NNCO. The new NNCO Deputy Director was hired in 2014 through an Intergovernmental Personnel Act mechanism, which enables temporary appointments to Federal positions from outside of the Federal Government, and the solicitation for the new NNCO Director welcomed applicants from industry and academia as well as Federal agencies.
Create a standing PCAST Nanotechnology Steering Committee of experts from industry, academia, and civil society to provide more frequent and in-depth guidance to the overall initiative and to the Signature Initiatives.	There is no evidence that a standing PCAST Steering Committee or other advisory body is planned.
Dedicate 0.3 percent of NNI funding to the NNCO to ensure the appropriate staffing and budget to effectively develop, monitor, and assess NNI programs.	The revised charter establishes a formula to determine the agency contributions to the NNCO budget. This funding approach is applied to the OMB nanotechnology cross-cut but does not comply with the 0.3 percent recommendation.
Work with the NNCO director to develop a plan for increasing the NNCO budget in line with its new responsibilities.	No new NNCO responsibilities are described in the revised NSET Charter or the 2014 NNI Strategic Plan.

The lack of progress against the 2012 recommendations is telling and disappointing. PCAST's 2012 review of the NNI provided recommendations concerning the leadership roles in NSET to ensure aggressive program management and appropriate representation from the broader government, academic, and commercial community engaged in the NNI. PCAST raised concerns that representatives appointed to the NSET Subcommittee might not have a level of authority within their agencies to influence budget allocations needed to meet NNI objectives.

In aggregate these recommendations aimed to build, within the recognized constraint that the NNI is not centrally funded, an NNI that is more aggressively focused (coordinated strategic plan, Signature Initiatives), more aggressively managed (Signature Initiatives, increased NNCO budget, NNCO as co-chair of NSET), and more broadly connected to the broad nanotechnology ecosystem (higher level agency representation on NSET, enabling NNCO director to be a non-Federal employee, creation of a Nanotechnology Steering Committee).

Building on this assessment from 2012 and with the additional focus on the transition to commercialization and the Grand-Challenges framework, we make a series of recommendations that increase the leadership role played by NSET and the Committee on Technology. We reiterate the concerns identified in 2012 that agency representation in NSET needs to be at a high enough level so that joint decision making can be accomplished and agency plans can be aligned to the direction coming from NSET strategic plan and, more importantly, the Grand Challenges. We find that the Committee on Technology, to which NSET reports, is not actively engaged in helping set and endorse priorities in nanotechnology and we strongly believe this must change. COT must be an integral part of the NNI. Finally, while many of these recommendations are not new, with the transition to NNI 2.0 and its increased focus on support for commercialization, we believe strongly that implementation of these recommendations is critical. If another two years passes without these kinds of changes, we believe the value of a centrally reported NNI will be substantially decreased.

Recommendation 4. NSET should continue to be co-chaired by the Assistant Director, Nanotechnology in OSTP and a representative of one of the participating agencies on a rotating basis. The Assistant Director should ensure that a leader for each active Grand Challenge also participates in a leadership role in NSET. NSET should use Grand Challenges as the primary vehicle for actively managing Federal nanotechnology activities toward directed outcomes

As required by the 21st Century Nanotechnology Research and Development Act of 2003, the National Nanotechnology Advisory Panel has the responsibility to carry out the biennial assessment of the NNI. By Executive Order this activity is housed in PCAST, and we believe this should continue. The 2012 report recommended that a "standing PCAST Nanotechnology Steering Committee of experts from industry, academia, and non-governmental organizations" be created "to provide more frequent and in-depth guidance to the overall initiative and to the signature initiatives."[4] We consider the underlying message of this unimplemented recommendation still to be valid. It is important that a broader set of expert voices from across the nanotechnology ecosystem be added to the conversations and deliberations of NSET, which is comprised solely of agency representatives. This becomes even more important within the context of the Grand Challenges framework as well as the expanded focus on transitioning research breakthroughs to commercial success. Input from industry, technology-transfer experts, and the venture-capital community become critical. We believe that a number of options exist, other than as a standing committee of PCAST, to create a group to provide such guidance and that is separate from PCAST's role in evaluating the NNI.

Finally, while we recognize that legislative action may be difficult at this time, we believe that aligning the separate PCAST review currently every two years and National Research Council (NRC) review currently every three year to the same concurrent three-year time interval can help reduce the burden on the NNCO, NSET, and agencies responsible for providing information.

Recommendation 5. PCAST should continue to carry out the Congressionally-required periodic review of the NNI. Congress should align this review and the triennial National Research Council reviews to the same concurrent three-year time interval to reduce the burden on the Initiative. OSTP, with the support of the NNCO, should create and administer a separate standing committee of cross-sector nanotechnology experts that provides guidance, but does not evaluate, the NNI.

Measurements

The processes and methods employed to measure the progress and return on the Nation's investment are critical to support the direction setting, prioritization, and leadership of the

NNI. Evaluation and tracking metrics have been the subject of recommendations in previous biennial reviews, and Table 5 provides an evaluation of the 2012 recommendations and our assessment of actions taken in response.

The Federal Government and the broader nanotechnology community would benefit from more rigorous and publicly available measures of the broad range of activities in the NNI, from discovery research to broad commercial deployment. To this end we provide recommendations for the ongoing research program and for the expanding commercialization activities. We note that these recommendations are also consistent with the Grand Challenge philosophy, which envisions not only clear measurements of ultimate success but also the establishment and tracking of progress to interim recommendations for each Grand Challenge.

Table 5. Response to 2012 PCAST recommendations on measurements

2012 NNI Review recommendation	Activities in response
Agencies should develop a mission-appropriate definition of nanotechnology that enables tracking specific nanotec hnology investments supported at the program level. The definition and funding details should be published in agency implementation plans to promote clarity.	NNI agencies utilize the NNI definition of nanotechnology and the Program Component Areas (PCAs) to track their investments and comply with the OMB cross-cut. The 2014 revision of the PCAs was developed, in part, to simplify program and investment tracking. Several or- ganizations, such as the Environmental Protec- tion Agency (EPA), Consumer Product Safety Commission, and the Food and Drug Admin- istration (FDA), have published mission- appropriate context to the definition of nano- technology to help them carry out their regula- tory responsibilities. There is no evidence that these contextual definitions change their in- vestment tracking at the program level.
The NNCO should track the development of metrics for quantifying the Federal nanotechnology portfolio and implement them to assess NNI outputs.	The NSET Subcommittee requested that the NRC provide advice on metrics in its 2013 review of the NNI. The NRC complied, but little responsive activity at the agency or NNI level has been observed. The 2014 Strategic Plan notes that the NNI agencies will continue to monitor metrics development that may be ap- plicable to nanotechnology. No NNCO role in tracking and implementing metrics is evident.

Measurement should capture activities at multiple levels of analysis and different temporal orders. It should also seek to describe the scientific enterprise in multi-level and potentially non-linear terms, not simply in terms of counts or trends but also in terms of complex and often multiple relationships among diverse system components. The current measurement of NNI influence is dominated by counts of publications and patents. This approach, while not unique to the NNI, is flawed for a variety of reasons, including mismeasurement of the process of science, cost, and burden[33,34,35]. In particular, patent counts do not measure commercialization or innovation, although they provide useful insights into the areas of activity. Measuring economic impact is much more complex[36,37]. The NSET with the Department of Commerce should begin a process now that will lead to useful measurements of nanotechnology activities to determine how the over $20 billion Federal investment in nanotechnology research has affected the United States economy. Establishing an accurate measurement system is not simple, but other institutions and organizations have already begun these kinds of measurement processes.

Universities and agencies – recognizing the need for accountability – are starting to build people-centered data systems to capture the impact of investments. One example is largely inspired by the project Science and Technology for America's Reinvestment: Measuring the EffecT of Research on Innovation, Competitiveness, and Science (STAR METRICS)[38]. In that system, data show the building blocks of research at *the project level*: showing the people who do the work and the firms that supply the scientific equipment[39]. The approach minimizes manual, burdensome reporting and uses existing data drawn from the human resource records and financial reports of universities[40]. The data have multiple advantages: they are low-cost, timely, broad-based, high-quality, and extensible. In addition, since they are based on administrative data from payroll and financial records, the source is immediately understandable to stakeholders[41].

This approach provides the foundation for a powerful new data infrastructure that could enable the integration of multiple sources of data and thus permit analysis of the role of funding in stimulating the creation, transmission, and adoption of ideas through those networks. In addition, it is possible to then develop more open and transparent measures of research activity and progress by capturing information on who is supported by research funding; what research is done, with whom and where; and what results flow from that research. We note that a data infrastructure that addresses these science policy goals is also likely to have much broader benefits, for example by enabling research on the sociology of science and by facilitating the navigation of knowledge networks by disciplinary researchers[42]. Appendix I, "Measuring the continuum of the research / translation process and subsequent outcomes using STAR METRICS," provides more details for a possible framework for measuring Federally funded nanotechnology research.

Recommendation 6. NSET, with the Department of Commerce, should execute a process to establish a common set of evaluation metrics to quantify and report the impact on workforce, productivity, and scientific knowledge in nanotechnology for all new research and commercialization programs beginning in FY2016

SECTION 5. COMMERCIALIZATION

Introduction

Fundamental nanotechnology research over the last decade has yielded an array of technologies that, over the long term, have the potential to transform energy, advanced materials, electronics, medicine, and more. This progress has brought the NNI to a point where the commercial potential that motivated the formation of the Initiative is within grasp. The U.S. nanotechnology enterprise, however, is not optimized to make practical use of nanotechnology advances. NSET and Federal agency leadership should go beyond coordinating basic research to ensuring a robust framework is in place, one that provides researchers who have promising laboratory advances the tools needed to attempt to transform their ideas into products that compete in the marketplace. Attention to commercialization in the nanotechnology community should not come at the expense of basic research; rather, the goal should be the development of a productive nanotechnology ecosystem that commercializes nano-based products through a sustainable pipeline fed by innovative basic research. The achievement of this broad strategic objective will not only ensure the introduction of novel, nano-based materials, medicines, and technologies, but it will enhance U.S. competitiveness and lead to infrastructure, capability, and work-force improvements – all areas of significant national interest.

Through NNI 2.0 the Federal Government should partner with universities, industry, and others to set up that ecosystem and to implement a series of Grand Challenges that set out clear technical goals and generate multi-sector investment in producing transformative competitive products. Section 3 described the process of Grand Challenges. This Section describes the current state of nanotechnology commercialization and summarizes the general technology-transfer process from lab to market. The specific hurdles to nanotechnology commercialization – reproducibly generating nanotechnologies and materials at high volume and addressing environmental, health, and safety (EHS) issues – are coupled with constraints common to all technology development: the skills of the entrepreneur, communication among stakeholders, and availability of capital. These hurdles lead us to three recommendations around manufacturing institutes, entrepreneur training, and forums for multi-sector engagement that are aimed to ensure the ecosystem is in place so that Grand Challenge technologies and other promising nanotechnologies reach the marketplace.

Commercializing Nanotechnology

The 2014 Lux Research report on nanotechnology describes a highly active industry sector in which commercialization of nano-based products continues to increase[22]:

- Revenues from nano-enabled products across all industry sectors grew to $731 billion in 2012.
- Total spending on nanotechnology research and development and start-ups bygovernments, corporations, and venture-capital (VC) investors reached $18.5 billion in 2012, an 8% increase over 2010, with $6.6 billion of this spending from the United States.
- U.S. Federal and State Governments provided $2.1 billion in nanotechnology funding in 2012.
- U.S. corporate spending on nanotechnology increased 19% from 2010 to 2011 and then 11% from 2011 to 2012, amounting to $4.1 billion in 2012.
- United States-based companies received the vast majority of VC investments, capturing 70% (a drop from 89% in 2010) of the $580 million invested in companies

developing nanotechnology in 2012. But very few Series-A investments (i.e., the first round of financing after seed capital) made in 2012 were attributed to the nanotechnology category, implying that either investors were focused on their current investments or nanotechnology-infused startups are being more broadly categorized as nanotechnology moves into more broad-based commercialization.

These aggregate numbers suggest that the substantial revenue generated from nano-enabled products and materials is approaching $1 trillion, investments by U.S. corporate and venture capital in commercializing nanotechnologies are relatively modest (approximately $5 billion in 2012). In the absence of metrics that would provide a more quantitative understanding of the nanotechnology ecosystem, it is difficult to draw firm conclusions about how efficiently new technologies are being commercialized, the return on investment in basic research, and whether current funding levels in the commercial sector can maintain a sustainable pipeline of high-value products over the long term.

In addition, these statistics provide little insight into the process, drivers, challenges, and stakeholders involved in moving a promising technology from the research laboratory toward the successful development of a product in the United States. Figure 8 illustrates one possible path that a promising technology might take when it translates from a fundamental technical advance to a marketed product. In the first stage, the academic entrepreneur is the key stakeholder for developing a new technology, and in the next stage, a start-up company might then take on the critical role of refining the promising laboratory discovery, validating its potential to yield useful products, and initiating the product development process[43]. It is important to note that the path as described here and in Figure 8 covers the transition from academic research to the market; other, more nonlinear paths are also common[44-48]. Similarly passages exist for innovation that starts in national laboratories and the basic research laboratories of commercial enterprises.

Federal agencies support entrepreneurs and start-ups through Small Business Innovation Research (SBIR) and Small Business Technology Transfer (STTR) programs, which awarded $110 million in nanotechnology grants across agencies in FY2012[16]. Public-private partnerships and precompetitive consortia are other mechanisms for enhancing commercialization. Examples include a number of public-private partnerships supported by the Semiconductor Research Corporation (SRC), which was established in 1982 as a consortium of U.S. semiconductor companies to fund and manage university research[49]. The SRC established the Nanoelectronics Research Initiative in 2005 to bring industry and academic researchers together to develop new electronic-switch technologies that enable substantial improvements in digital performance.

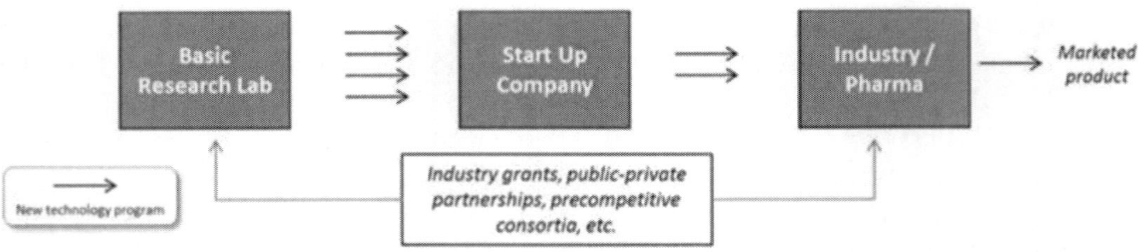

	Academic Research Lab	Start Up Venture	Pharmaceutical Industry
Key Stakeholders	Academic researchers Funding agencies Grant peer-review panels	Academic entrepreneurs Venture capitalists (VCs) Strategic investors (industry venture funds) SBIR/STTR	Technology companies Pharma / large biotech Regulatory agencies (e.g., FDA, EPA)
Ecosystem Constraints	Pursue 'fundable' projects 'White-space' projects may face high funding hurdles Limited view of industry technology gaps & needs Relatively few successful academic entrepreneurs Poor understanding of commercialization process	Strong VC preference to work with established/ successful academic entrepreneurs VCs now favoring later stage funding rounds to advance established companies Start-up funding may be difficult to obtain	Limited view of emerging technologies in academia Pharma relies heavily on biotechs – limited in-house nano activities Seek nanotech solution as a last resort Lack of established EHS standards, practices, and metrics to guide nano commercialization
Infrastructure and Workforce Constraints	Stability of national nanotechnology centers Appreciation for entrepreneurial skills needed to commercialize basic science	Must develop/build fabrication capabilities for early technology development Scale up for advanced testing Cost prohibitive access to some user facilities for proprietary work	Large capital investment may be needed to produce scaled-up quantities for product testing/ validation Nanomanufacturing facilities not readily available GMP facilities must be available early in the clinical development process for drugs. Need to invest before drug enters clinical development Incentive to go outside United States
Funding	Basic research constitutes majority of U.S. nanotechnology spending Numerous agencies fund basic nanotechnology research under the NNI umbrella United States lacks a well-defined nano strategy	VC funding represents a small percentage of the national nanotechnology spending Public funding of nano commercialization projects (e.g., SBIR) cannot fully support nano start-up efforts Industry support for advancing select nano programs through licensing, collaborations, or strategic investments	Industry typically bears the cost of product commercialization, including building the supply chain Build new manufacturing capabilities internally or work with CMOs (United States or abroad)

Figure 8. Commercialization of technology from academic laboratory to the pharmaceutical market.

Constraints Limiting the Commercialization of Nanotechnology in the United States

Although the commercialization model shown in Figure 8 involves well known stakeholders at each stage of the process, a number of constraints inhibit nanotechnology product development. Peer-review of high-risk, high-return ideas, inexperience of first-time entrepreneurs, limited communication among stakeholders, and limited capital for new entrepreneurs are stumbling blocks to commercializing any technology (Box 7). But in nanotechnology, regardless of whether it is a nano-based advanced material, a targeted nanoparticle with a drug payload, or a nano-based diagnostic tracer, generating volumes of nanoscale material in a reliable and consistent manner is required for product testing, validation, and demonstration of scale-up potential for manufacturing.

Alleviating two constraints on nanotechnology commercialization, in particular, could enhance the nanotechnology ecosystem:

- *Investment in nanofabrication facilities* – The commercialization of nanotechnology innovations depends heavily on the successful development of nanofabrication and nanomanufacturing procedures. Few nanomanufacturing user facilities are accessible for developing production procedures, scaling up volumes of nanomaterial for research, or generating commercial supply. In the absence of these facilities, start-ups must assume significant up-front financial risk in developing in-house facilities to support company operations. In addition, the production of nanomedicines for clinical testing and commercial supply must be done in a nanomanufacturing environment that meets the Food and Drug Administration's (FDA) requirements for Good Manufacturing Practice (GMP). Making a significant financial investment in building and qualifying GMP facilities is a considerable risk for a company before the nanomedicine has been assessed for efficacy in human clinical trials to support FDA approval.
- *Comprehensive nanotechnology EHS standards* – Notwithstanding recommendations from previous NNI review panels, no comprehensive EHS standards, best practices, or metrics have yet been established to guide the commercialization and manufacture of nanomaterials. With the continued introduction of nano-based products across industries, some potentially manufactured at a large scale, adequate exploration and quantification of potential risks to safety and the environment and the development of best practices to mitigate any true risks that are identified are needed.

Box 7. General Constraints to Technology Commercialization

- *First-time academic entrepreneurs* – Many faculty members have little training or experience in moving a technical innovation out of the research laboratory into a small company, and the level of support within a university can affect a researcher's ability to spin-off or license a technology. Critical issues related to intellectual property, strategies in acquiring a technology license from the university, forming a company, and acquiring funding all need to be addressed in the early stages of spinning out a start-up company. Unless the new entrepreneur has access to mentors and other resources, promising innovations may not get out of the laboratory.
- *Communication among stakeholders* – Successful academic investigators understand the technical landscape and the potential value of their work, but they may not know how their innovations might address strategic gaps at a large technology company or could be translated into a commercial success. A company R&D director, conversely, might know little about a high-value technology being developed in an academic laboratory. Additional venues are needed to bring together academic entrepreneurs, VCs, industry, relevant Government agencies, and other stakeholders.

- *Venture capitalfor new entrepreneurs* – Top VCs tend to work with successful academic entrepreneurs with proven track records because VCs perceive these individuals have to lower investment risk. This preference, however, makes fundraising more difficult for the lesser known or less experienced entrepreneur. In addition, the current trend is for VCs to invest in later-stage funding rounds at established companies rather than in early-stage start-ups.
- *Peer review of high-risk, high-return ideas* – Academic investigators pursuing potentially transformative high-risk, high-return projects may find it difficult to obtain funding because of a conservative peer review system. A reviewer may not recommend funding for a proposal because of the perceived risk of the idea or the perception that the work may not align with the granting agency's objectives or risk tolerance.

Of the commercialization constraints specific to nanotechnology – nanofabrication facilities and EHS standards – the first is related to the development of nanomanufacturing in the United States. The latter will be discussed in Section 7. A recent Government Accountability Office (GAO) report focused on nanomanufacturing and the effect that nanomanufacturing capabilities have on U.S. competitiveness[7]. The report emphasized that, while the United States remains preeminent in nanotechnology research, it has fallen behind other countries in developing the necessary nanotechnology infrastructure and workforce. The major impediment in this commercial translation typically occurs in the "Valley of Death" or the "Missing Middle," the stage where infrastructure is needed between Government-backed research and select privately-funded technology scale ups (Figure 9).

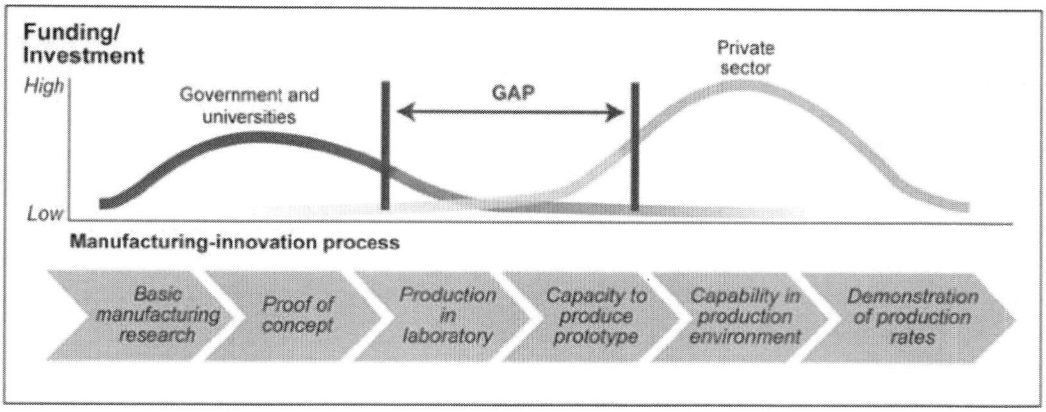

Figure 9. Early stage technology development is typically supported by grants to academic investigators, and late stage private sector investments are made to select technologies with significant potential, leaving a gap in the middle. Figure adapted from GAO report[7].

The United States has lost leadership in several industries to other countries that have emerged as strong manufacturing centers. Semiconductors, advanced batteries for vehicles, and cement-based construction materials, all of which were originally developed in the United States, are now manufactured elsewhere. These examples illustrate how an early competitive advantage in developing high-value technologies can be lost if commercialization is not adequately supported through the development of the necessary manufacturing infrastructure[50-52]. Exciting advances have recently been made in the field of nanomedicine and nanobiology that could enable new approaches to disease treatment, but the United States risks its leadership in this area without a proper ecosystem to support it. Additional details on manufacturing nanomedicine are in Appendix II.

NSET formed the Signature Initiative on Sustainable Nanomanufacturing - Creating the Industries of the Future to spotlight Federal coordination in this area, and in a July 2010 white paper, the Subcommittee describes thrust areas coordinating around the design of scalable and sustainable nanomaterials and processes and around nanomanufacturing measurement technologies, setting out two-year, four-year, and eight-year goals. The FY2014 estimated spending across agencies on nanomanufacturing was $38.4 million, and the FY 2015 budget request is $36.2 million: $6.1 million for National Institute of Standards and Technology (NIST), $0.8 million for Department of Defense, $3.9 million for National Aeronautics and Space Administration (NASA), $23.4 million for National Science Foundation (NSF), and $2.0 million for Department of Agriculture (USDA).

The Department of Health and Human Services, which invested an estimated $469.5 million across all its nanotechnology portfolio in FY2014, estimated no funding for nanomanufacturing in FY2015 but did report the National Cancer Institute (NCI) formed the Translation of Nanotechnology in Cancer consortium to facilitate communication among patient advocates, the Agency, and the pharmaceutical/biotechnology industry and also funded the NCI Nanotechnology Characterization Lab. The NSF funds four Nano Science and Engineering Research Centers (NSECs) that focus on nanomanufacturing; two centers will retire in 2014 and the other two will retire in 2015 and 2016. The four NSECs in partnership with NIST's Center for Nanoscale Science and Technology, the Center for Integrated Nanotechnologies at Sandia and Los Alamos National Laboratories from the Department of Energy (DOE), and other affiliates together form the National Network for Nanomanufacturing, which organizes an annual nanomanufacturing workshop. In 2014 the USDA began a larger effort to explore the manufacture of nanocellulose materials.

While each agency considers areas of nanotechnology interest according to its mission, the challenges in manufacturing are likely to impede commercializing advanced nanomaterials, nanomedicine, and other nanotechnologies unless the Federal Government addresses the valley of death, which involves the need for nanofabrication facilities to create high volumes of nanotechnology product. Engaging the broad manufacturing initiative established by the Obama Administration is a promising approach.

Manufacturing Innovation Institutes Dedicated to Key Nanoscience Areas

In 2012 and building on past PCAST recommendations, the Administration announced plans to create a National Network for Manufacturing Innovation (NNMI) "consisting of regional hubs that will accelerate development and adoption of cutting-edge manufacturing technologies for making new, globally competitive products." These hubs, called Manufacturing Innovation Institutes (MIIs), are intended to strengthen U.S. competitiveness, initiate new ventures, and boost local and state economies – with the broad strategic goal of innovating and scaling up advanced manufacturing technologies and processes[53]. Importantly, the institutes are to be industry-led. They will require strong partnerships among industry (companies large and small, established and start-up), academia, non-profit organizations, and states, with the aim of investing in and accelerating the development of cutting-edge manufacturing technologies.

To date four MII's have been established and a fifth in competition[54]. In both his 2013 and 2014 State of the Union addresses, the President asked Congress to authorize a $1 billion investment to create a network of up to 15 regional MMIs. The NNMI initiative has significant potential to address a critical challenge now being faced by companies trying to advance nano-based products, nanomedicines, and advanced nanomaterials.

Section 3 outlined the characteristics and best practice processes to choose a Grand Challenge. Given the strong worldwide competition to achieve the best nanomanufacturing capabilities, we believe at least one of the first Grand Challenge selected should be mature enough to take on manufacturing challenges as the results will be needed to achieve almost any Grand Challenge that comes after it. And we believe a good place to start is to leverage the current Manufacturing Innovation Institutes that have been created under NNMI.

Recommendation 7. NSET should work with Federal agencies to define potential Manufacturing Innovation Institutes dedicated to nanoscience and nanotechnology as part of the National Network for Manufacturing Innovation program.

I-Corps Training for New Entrepreneurs

Another constraint on commercialization, which is common to many technology areas, is that training for a scientific or engineering career typically does not provide a working knowledge of finding start-up funding, obtaining technology rights, developing a business strategy, and running a company, all skills helpful if not essential for moving high-value technologies out of the research lab. To begin to address this situation, the National Science Foundation introduced the NSF Innovation Corps (I-Corps) program. Its goal is to prepare scientists and engineers to "extend their focus beyond the laboratory and broadens the impact of select, NSF-funded, basic-research projects." I-Corps Teams are composed of academic researchers, student entrepreneurs, and business mentors who pursue a curriculum that provides real-world, hands-on learning about what it takes to successfully translate fundamental technical advancements into commercial products[55].

Importantly, the I-Corps program is not just training but intends to foster entrepreneurship that directly leads to the commercialization of a technology that has been supported previously by NSF-funded research. Program-commercialization activities will most often focus on start-ups founded by the I-Corps participants; successful I-Corps projects are ready to form businesses. The NSF SBIR and STTR programs are important vehicles for launching those businesses.

Currently only NSF grant awardees can apply to participate in I-Corps, but we see strong value in expanding this program across other funding agencies (e.g., National Institutes of Health {NIH} and the Department of Energy) and significantly enhancing the innovation landscape across technology sectors, including nanotechnology. PCAST supported the development of the I-Corps program in our 2012 report, *Transformation and Opportunity: The Future of the U.S. Research Enterprise*[56]. Only a few of the nearly 300 NSF I-Corps grantees have been working on nanotechnology projects. The NSET and agencies could work with NSF to support an I-Corps hub in nanotechnology that accepts teams funded from any agency on successful nanotechnology projects.

Recommendation 8. The NSF, in consultation with NSET, should expand the NSF Innovation Corps to include a specific focus on entrepreneurship in the nanotechnology area, and NSET, NIH, and DOE should leverage this program concept to make it available to a broad range of scientists working in nanoscience and technology.

Robust Communication across the Nanotechnology Ecosystem

As occurs in many industries, poor communication among stakeholders impedes the ecosystem for nanotechnology commercialization. As discussed above, the new academic entrepreneur may not know technology gaps that large technology or pharmaceutical companies face, and companies likely do not know the value of new scientific findings from academic laboratories. Some venture-capital fund managers might prefer to work with established investigators who are experienced entrepreneurs because they decrease the investment risk, while the lesser-known, new entrepreneur may have a difficult time accessing funding for an innovative finding.

To address some of these issues, organizers of a variety of national scientific meetings have established partnering sessions for academic investigators, start-ups, and large companies to explore their mutual interests during major meetings. The National SBIR/STTR Conference is one avenue that crosses technology areas. Providing these partnering meeting opportunities at appropriate national meetings with nanotechnology interests could enhance the nano ecosystem in ways that further enable the commercialization of nanotechnology products. The annual Nanotechnology Conference and Expo, most recently as Nanotech 2014 organized by the Nano Science and Technology Institute, begins to address some of these communication gaps. It began in 1998 and is now co-located with several other industry-specific technology innovation expos. The annual Nano Technology for Defense conference hosts partnering meetings for those in the defense industry. With new entrepreneurs in nanotechnology arising from many different disciplinary fields, evaluating whether additional partnering sessions at disciplinary- or industry-specific conferences would be valuable. The Department of Commerce should leverage its experience with other technology industries and the support of the National Nanotechnology Coordinating Office to create a regular economic-development forum that focuses on commercial development. With nanotechnology research alive in so many different fields, the nanotechnology community should ask what additional measures are needed to fully engage the range of stakeholders toward any one nanotechnology Grand Challenge.

Along similar lines, information about nanomaterials could bring commercial and research activities together. Already NanoHUB.org provides online tools for researchers and more to teach, learn, and collaborate around nanotechnology, and the Nanomaterial Registry curates data about nanomaterials' interaction with biology and the environment. Establishing a *'clearing house' for information* related to the general nature of nanomaterials disclosed to be in the development process (non-proprietary information) could create a resource for exploring available nanotechnology solutions to key technical problems being faced across various industries. Assessing how to communicate effectively with industry with a focus on identifying opportunities and fostering technology translation will be important in NNI2.0.

Finally, a key limitation that the nanotechnology ecosystem suffers from is the current inability to measure accurately the economic impact that nanotechnology commercialization has had. Nanotechnology has had a significant impact on multiple industry sectors; the role that nanotechnology has played, however, is not always obvious and the development of robust metrics that quantify the commercial impact of nanotechnology could provide strong support for continued investment in this key technology area and highlight areas that should be supported in the future.

Recommendation 9. The NNCO and the Department of Commerce should establish an annual nano-focused economic-development forum designed to bring together academic researchers, the venture capital community, biotechnology, and other industry in a format that enhances the possibility to create business partnerships.

Identify and Incentivize University, Local, State, and Regional Best Practices That Encourage Technology Commercialization

Training in business skills and development will help first-time academic entrepreneurs, but the resources available to a first-time academic entrepreneur within her/his own university or institute may equally affect whether a research advance leaves the laboratory. The resources and culture for commercializing research vary significantly across U.S. universities and national laboratories. Patenting and technology transfer cannot take place without institutional support. The institution plays a critical role in guiding the patent application process, providing funding for the costly prosecution of patent applications and enabling the licensing of technologies to outside parties. Institutional support for technology translation can be enhanced through university administrations placing a strong focus on patenting important findings of their researchers, by facilitating technology and material transfer among collaborators, and establishing business development offices[57]. Due to the high cost and time involved, researchers often publish their work without filing strong patent applications[58]. In doing this they run the risk of losing exclusive access to their own technology, making it very difficult for them to commercialize, although potentially creating positive effects on the market as a whole[59]. VCs, strategic investors, and large companies are unlikely to invest in technologies that do not have a strong proprietary intellectual property position. By providing modest funding and legal assistance, the home institution can significantly lower the barrier to patent filing and significantly increase the probability that the technology can ultimately achieve strong patent protection.

Among the universities with active technology transfer offices, the range of support likely varies widely. Building upon the National Research Council study *Managing University Intellectual Property in the Public Interest*, an appropriate expert group potentially supported in part by the NNI agencies should survey a range of institutions that have been successful in encouraging technology transfer to the private sector to determine best practices in the following areas: (i) amount spent on tech transfer per \$100 million in research dollars; (ii) number of technology transfer staff per \$100 million in research dollars; (iii) conflict-of-interest management; and (iv) responsibility of researchers in submitting disclosures[60]. This information may lead to a more efficient use of Federal funding that could result in the enhanced translation of NNI funded research. The survey should also explore practices that universities and national laboratories have in place regarding the publication of key data prior to filing the appropriate patent applications. NNI agencies should also design and implement a plan that motivates a stronger emphasis on the commercialization of relevant technologies that emerge from institutions involved in Federally funded research.

In addition to support at universities, State and regional activities have fostered nanotechnology commercialization. Regional, State, and local nanotechnology initiatives have existed in at least 24 States to provide resources and facilitate collaboration between the private sector and academic researchers. These initiatives were supported with combined funding from Federal, state, local, and private partnerships. Due to the youth and volatility of nanotechnology research in conjunction with high-risk business models, most of the initiatives have diminished or completely shut down because of their inability to retain strong Federal and State support, especially through the recession[61]. The continued influence of initiatives that have ended suggest that the sustainability of nanotechnology initiatives may be less important than their ability to have an immediate impact on the prevailing state of nanotechnology research. While many initiatives have come to a close, thriving start-up and small and medium enterprises have been built as a result of the programs[62].

We note two specific examples. The Oregon Nanoscience and Microtechnology Institute (ONAMI) is a State government investment model that supports researchers and small businesses by encouraging collaborative research with matching grants, supporting access to shared laboratory spaces, and offering funding and expertise for commercialization. Since its inception in 2004, ONAMI has helped produce 32 startup companies and 61 patents. The Arizona Nanotechnology Cluster is an example of a volunteer and local business funded initiative. Since 2003, the Arizona Nanotechnology Cluster has provided seminars, networking, and educational opportunities, and has served a voice for Arizona nanotechnology in local, State, and national government. The Cluster partners with State research universities and nanotechnology businesses to increase the role of nanotechnology in the Arizona economy.

SECTION 6. THE NNI RESEARCH ENTERPRISE

Introduction

In NNI 2.0, Federal agencies, NSET, and the NNCO need to focus on setting up platforms from which nanotechnology can launch into commercial products, but as those efforts build, continued support of the research enterprise must ensure that discovery and exploratory research remains vibrant. As the NNI matures, achieving Grand Challenges will require a balanced portfolio of activities by the NSET and Federal agencies in discovery, exploration, and commercialization that will enable the United States to simultaneously reap the benefits of past NNI investments and discoveries, while ensuring that it is planting the seeds of discovery that can be harvested indefinitely into the future. *Without discovery research now, there will be nothing to commercialize in the future.*

The investments of the National Nanotechnology Initiative have already yielded tremendous nanoscale science in the United States. The first 10 years of the NNI was a period when researchers built the nanoscience foundation for technology – creating, studying, and understanding new classes of nanoscale building blocks. In the next phase of research, the goal is to build on that foundation by controlling the structure of those building blocks reproducibly and at the atomic scale, by moving from passive to active nanostructures, and by integrating nanoscale elements into complex assemblies with emergent and, ultimately, predictable properties. The understanding required to bring these to fruition under NNI 2.0 will be critical to solving many of the major scientific mysteries and societal challenges facing the United States. Achieving this control, function, and integration, however, is more difficult than the creation of the nanoscale units themselves. As in the path to electronics, the discovery of the transistor was only the first step in a long journey to laptops, smartphones, and the Internet.

The path from discovery to innovation has three distinct phases, as shown in Figure 10: discovery, exploration, and commercialization. Each stage requires different resources and focus, but all three are essential. This Section focuses on actions to support the first two steps, discovery and subsequent exploration.

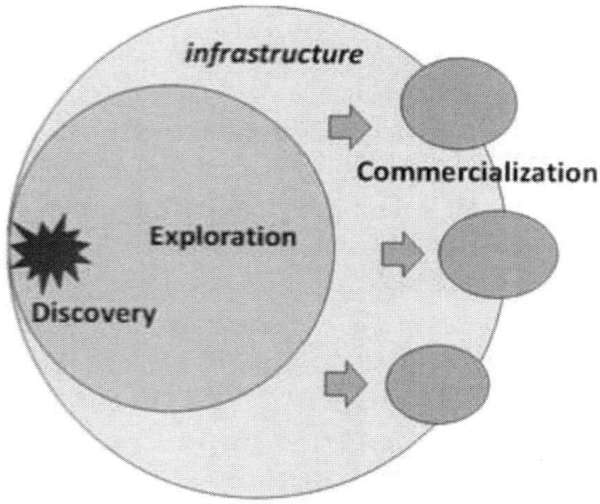

Figure 10. The stages of technology development begin with basic research discovery (inner star). In the discovery phase, new science or technology happens and high-risk, high-reward activities occur. During the exploration of these discoveries (medium green circle), their implications are investigated and their potential impact can begin to be assessed. Finally, with the support of infrastructure (light green circle), commercialization stage occurs when the discovery is incorporated into commercial applications (dark green circles).

Major research breakthroughs are needed during NNI 2.0, including:

- Atomic scale control – new approaches to create complex nanostructures with the atomic-scale precision of their biological equivalents.
- Active nanostructures – creating nanostructures that are active, i.e., perform a complex physical, chemical, or electronic function.

- Integration and emergence in nanoscale assemblies – creating assemblies of nanostructures whose behavior emerges from the collective and is not straightforwardly predicted from the behavior of the individual constituents.

PCAST believes that the most successful approach to achieving these goals is to focus on not just what but *how* interdisciplinary science gets done during discovery and exploration. In particular, the recommendations for the *discovery* phase will empower **people** to do their most creative work to generate new breakthroughs, and recommendations for the *exploration* phase will **focus** the efforts of large groups on specific objectives as the goals become clear. Federal agencies should continue to support fundamental research in nanoscience and nanotechnology.

Investing in Creative, High-Risk Research Through Single Investigators

New ideas and paradigms emerge when creative minds are given free rein to explore. By definition, breakthroughs cannot be anticipated in detail or planned in advance, but creating the right environment allows them to flourish. And new ideas are the seeds from which future technologies grow. Despite this case, a crisis is growing in American nanoscience: top researchers are leaving the United States for senior positions around the world[63,64]. Researchers are leaving because of the challenge in obtaining flexible, long-term funding that supports their ability to generate and sustain their most creative ideas. Stopping this trend is essential. The cultivation of top researchers and their creative ideas cannot be neglected without severely damaging the long term prospects of nanotechnology leadership by the United States.

We believe that the best way to foster discovery is to identify the country's most talented people and empower them to explore their vision. Providing highly skilled and talented people with the resources to take risks and explore new directions needs to occur at every stage of the science and technology researcher's career: at the graduate level, at the new faculty level, and at the senior faculty level. The Federal Government supports a number of fellowship programs for graduate students, and programs such as the Young Investigator Programs by the Office of Naval Research, the Army Research Office, and the Air Force Office of Scientific Research and the Presidential Early Career Award for Scientists and Engineering by the National Science Foundation (NSF) support new faculty. Private foundations grant awards such as the Packard Fellowships and Sloan Fellowships to new faculty as well.

These types of support launch new faculty careers, and equally important, they encourage risk-taking and discovery at what is often the most creative and paradigm-shifting stage of a researcher's career. New faculty have new ideas, and empowering them at this stage to explore their scientific vision is a powerful force for change and for breakthrough discovery. PCAST encourages the NSET agencies to continue to provide strong research support to outstanding early career researchers in nanoscience and technology to pursue what the researchers consider to be the most promising avenues of research.

For senior researchers, the story is different. Top researchers in the United States are leaving for either Europe or Asia because they cannot secure long-term, stable funding in the United States for their most visionary work. The U.S. funding system is fragmented among agencies beholden to individual missions and is captive to the peer-review system that not only appropriately weeds out the inferior ideas but also potentially dismisses truly revolutionary ones. This perfect storm makes it nearly impossible for a faculty member to invest the time and effort both to obtain grants that support a significant research group and to pursue her or his most creative, high-risk ideas. Nanoscience and technology is receiving world-wide interest and research investment, yet even the very best researchers suffer from the vagaries of funding that limit their creativity, productivity, and ability to pursue the most promising, but high-risk, areas of nanoscience and technology research. This uncertainty in raising funding is eroding U.S. competitiveness in retaining the best scientists in the world.

European and Asian competitors offer very attractive positions to address the needs of top, senior researchers. Positions such as the Max Planck Institute Directorships in Germany or the ERC Pioneer Program in Europe provide a strong, consistent base of funding that allows the best researchers to explore high-risk, high-reward science. The United States needs funding opportunities to support the Nation's best and most productive researchers in a steady and substantial manner that allows them to concentrate on developing creative ideas. A model is the National Security Science and Engineering Faculty Fellowship (NSSEFF) program at the Department of Defense (DOD). This highly competitive program supports faculty with approximately $3 million over five years to do cutting-edge science that has significant national security implications. The National Institutes of Health (NIH) similarly awards highly competitive Pioneer Awards at $500,000 per year for five years for high-risk research by a single investigator at any stage in their faculty career.

We propose a Nanoscience Faculty Fellowship Program modeled on the NSSEFF or as an extension of that program to support the most innovative and creative researchers in nanotechnology. These researchers will also serve as a valuable pool that could be tapped for oversight, review, advice, and management of the NNI portfolio in a manner similar to what occurs with the NSSEFF for national security issues. Approximately five substantial awards per year that each last five years would result in steady-state support for 25 nanoscience researchers. We believe this is a highly effective means to support creative, high-risk research, enabling highly talented and productive researchers to pursue work not likely to be funded by traditional mechanisms and encourage them to stay in the United States.

Recommendation 10. Recognizing growing international funding competition that is attracting U.S.-based talent to go abroad, NNI agencies should substantially support the best single investigators to pursue creative, high-risk research. In particular, the NSF, DOE, DOD, and NIH should coordinate to ensure that at least five new National Security Science and Engineering Faculty Fellowship (NSSEFF)-style senior investigator grants in nanoscience and nanotechnology are funded per year.

Exploratory Research Through Research Centers

Achieving some Grand Challenges in nanotechnology will require transdisciplinary, exploratory research, and Federal agencies can catalyze that exploration by investing in research centers. Centers can be teams of co-funded investigators collaborating on a common research vision using investigators' own labs – either at one institution or distributed across many. Centers can also be critical physical laboratories that provide the proper research infrastructure and expertise for nanotechnology science and technology that are essential components of the NNI mission. Centers in either case provide the research environment and the training of next generation scientists and engineers who are needed to achieve Grand Challenges.

Transdisciplinary Research

Federal agencies support a range of research centers that can be forums of collaboration for transdisciplinary research in nanoscience. National Cancer Institute's (NCI) Centers for Cancer Nanotechnology Excellence and NSF's Nanoscale Science and Engineering Centers focus on nanotechnology alone while NSF's Engineering Research Centers, DOD's Multidisciplinary University Research Initiative, DOE's Energy Frontier Research Centers, and NSF's Materials Research Science and Engineering Centers are successful programs with broad portfolios that support nanotechnology research. Nanotechnology-focused centers can also be as large as NSF Science and Technology Centers or DOE Energy Innovation Hubs. We believe research centers are critical for the exploratory research that will translate discovery research into technology.

While research centers offer myriad potential benefits, they have key risks. In particular, the cost of funding unfruitful research is much greater in a multi-investigator research center than in an individual investigator award. If the research of a single investigator is unsuccessful, a considerably smaller amount of funds are lost in comparison to the failure of a center to produce. One way to mitigate this risk is to ensure that each center has strong leadership that will be able to guide its path away from unrewarding efforts, toward meaningful and productive research. Ensuring strong leadership at the helm of centers may be done in several ways. One technique used by the DOE in managing the Energy Frontier Research Center program is to embed a management review after the first year of a center's funding. A mechanism DOD employs in centers is to utilize pilot grants for centers, whereby continued funding is contingent on a strong start in the first year of the center.

Another pitfall of center-based research is the potential to stifle innovation inadvertently through an overly rigid structure of reporting and requirements. To avoid this, agencies could outline broad goals for a successful center rather than a strict framework of research requirements. This will allow ideas to "bubble up" from investigators and allow researchers to pivot from one project to another should an unexpected, promising discovery be made.

In order to maintain the health of a research center ecosystem, centers should have built-in checks and balances to ensure progress is still being made. This will allow agencies to shut down unproductive centers that repeatedly fail to meet goals while continuously funding centers that are doing well, regardless of an arbitrary granting timeline.

Center-based grants are an important part of the any research portfolio, especially in nanoscience and nanotechnology. To train students who can think critically about the problems facing society today, especially in the realms of nanoscience and nanotechnology, program managers and faculty must find ways to overcome disciplinary constructs to train a new type of scientist: one who is not only "multilingual" across multiple scientific fields, but who can speak a common language and foster communication and collaboration among many disciplines.

The Grand Challenges discussed in Section 3 will be a powerful focal point for exploratory research. We propose that the Federal agencies support a program of nanoscale-focused research centers. One example could be a center supported by NSF, DOE, and

National Institute of Standards and Technology (NIST) as a Particle Foundry, which would provide the research community a unique educational and training resource for the manufacture and production of nanoparticles by batch, continuous, or microfluidic methodologies. Additional nanoscale-focused research centers may be supported by a single agency or in a joint effort by several agencies, but these nanoscale science and engineering centers should be funded at least 50% by a single directorate or agency that functions as its disciplinary home. As an exemplar for the type of centers to be established, agencies can look to the NCI and its robust Alliance for Nanotechnology in Cancer initiative, including the Nanotechnology Characterization Laboratory. Further value can be added using this model even within the NIH by instituting a center program analogous to the NCI to address disease indications other than cancer including autoimmune diseases, infectious diseases, and more.

Infrastructure

DOE's Nanoscale Science Research Centers, NCI's Nanotechnology Characterization Laboratory, NIST's Center for Nanoscale Science & Technology, and NSF's National Nanotechnology Infrastructure Network (NNIN) provide essential infrastructure for nanotechnology discovery and exploration. They provide ready access to specialized tools that are generally too expensive for each laboratory, institution, or company to acquire. They provide essential training in interdisciplinary nanoscale approaches and techniques to new generations of researchers, industrial engineers, and entrepreneurs.

The NNIN is a network of 14 university-based user facilities that provide open-access tools and resources to nanotechnology researchers in universities, industry, and Government. Established in 2004 through a 10-year NSF grant awarding a total of $179.8 million, the NNIN provides over 1100 advanced instruments at affordable costs to "enable rapid advancements in nanoscale science, technology, and engineering."[65] The NNIN also maintains educational and outreach opportunities through initiatives such as the Research Experience for Undergraduates program.

With the original NNIN set to expire in 2014, the NSF is developing a successor program in FY2015. As the existing infrastructure retires and the agency begins a new program, providing open access, accessible facilities that reach a very wide audience in a cost-effective manner, both in the public and private sectors will be critical. The retiring NNIN labs are strongly leveraged, with NNIN funding only a small, but key, part of the overall budget. In particular, the NNIN funding helps to link facilities across the country together and keep them open to outside users. Without the NNIN funding, many of these facilities likely would be significantly less accommodating to outside users, and in case of the smaller participants, might no longer be viable at all.

Recommendation 11. NSF, NIH, DOE, DOD, and NIST should strongly support nanoscale research centers and infrastructure networks to ensure the effective training of a new generation of transdisciplinary scientists and engineers, in particular by strongly supporting the Next-Generation National Nanotechnology Infrastructure Network.

SECTION 7. ENVIRONMENTAL, HEALTH, AND SAFETY ISSUES

Introduction

PCAST's *Fourth Assessment of the National Nanotechnology Initiative* lauded the "significant progress" made by the NNI to address potential environmental, health, and safety (EHS) risks of nanotechnology[4]. It noted the rapid growth of research funding for nanotechnology EHS issues, the implementation of PCAST's earlier recommendation to identify a central coordinator for EHS-research within the NNCO, and the expansion of the Nanotechnology Environmental and Health Implications (NEHI) Working Group charter to address cross-agency nanotechnology related policy issues. The fourth assessment praised the development and release of an EHS research strategy that incorporates nanomaterial measurement capabilities, human exposure assessment, human health, environment, risk assessment, risk management, and predictive modeling and informatics. It also called out significant NNI progress in addressing the potential EHS risks of nanotechnology, noting the development and release of a national EHS research strategy and articulating an approach that incorporates the "evolving research needs and the strategic research plans of three relevant agencies."

While we noted that there was good agreement between the NNI strategy and the findings of the 2012 National Research Council (NRC) report *A Research Strategy for Environmental, Health, and Safety Aspects of Engineered Nanomaterials*, we also expressed concern about the "lack of integration between nanotechnology related EHS research funded through the NNI and the kind of information policymakers need to effectively manage potential risks from nanomaterials."[66] This integration is important to promote commercialization and obtain regulatory approval for products to reach the marketplace. This sentiment was echoed by the 2014 NRC report *Research Progress on Environmental Health and Safety Aspects of Engineered Nanomaterials*, which concluded that risk characterization by engineered nanomaterials (ENMs) "across their life cycles is a scientific challenge that requires integrated, quantitative and systems-level scientific approaches."[67] The latter report includes a detailed outline of the research progress since 2012, classifying the progress as substantial, moderate, or minimal. In this assessment, we will focus our recommendations on the integration of knowledge and research to improve regulatory decision making and commercialization of nanotechnology with the acceptance, trust, and support of the public.

Review of 2012 PCAST Recommendations

Table 6 shows a summary of the recommendations made in our 2012 NNI review along with an assessment of actions taken against those recommendations. We find good alignment and progress between the recommendations and the actions.

Table 6. Response to 2012 PCAST recommendations on environmental, health, and safety issues

2012 NNI Review recommendation	Activities in response
Establish high level, cross-agency authoritative and accountable governance of Federal nanotechnology related EHS research so that the knowledge created as a result of Federal investments can better inform policy makers.	The NSET NEHI Working Group remains a very active and collaborative cross-agency forum that released a review of NNI EHS research activities, accomplishments, and collaborations, *Progress Review on the Coordinated Implemen- tation of the NNI 2011 EHS Research Strategy* (June 2014). Several NEHI agencies also partici- pate in the Emerging Technologies Interagency Policy Committee Nanotechnology Working Group and provide information on their re- search activities directly to policy makers.
Increase investment in cross-cutting areas of EHS that promote knowledge transfer such as informatics, partnerships, and instrumentation development.	The Nanotechnology Signature Initiative (NSI) on Nanotechnology Knowledge Infrastructure will provide tools and empower research communities to translate EHS data to information and knowledge. Investment in this NSI jumped from $7.2 million (actuals, FY2013) to $32.1 million (estimate, FY2014) and $26.2 million (request, FY2015). No investment breakdown for partnerships and instrumentation is availa- ble. The EHS budget remains steady at $115 million (actuals, FY2013), $113 million (esti- mate, FY2014) and dips slightly to $112 million (requested, FY2015). These figures include in- strumentation and partnership activities. Part- nership activities, defined as Centers and Net- works, are identified in the NNI EHS Progress Review, Appendix C.

Current Status of Nanotechnology EHS

The current status of nanotechnology EHS is characterized by (i) a reliance on hazard and risk analysis on a case-by case basis in spite of the rapid expansion of the technology; (ii) a paucity of exposure, dosimetry, and lifecycle data because of the lack of information about materials in the commercial chain, as well as slow emergence of the instrumentation required to detect ENM exposure under complex biological and environmental conditions; (iii) continued reliance on descriptive (non-mechanistic) animal studies (often poorly integrated with cellular and *in vitro* data) for regulatory decision making; and (iv) regulations based on ENMs as novel chemical substances which, by way of new use rules and pre-manufacturing notices, negotiate access to marketplace without a coordinated approach between Government agencies, industry, and academia[68]. At the same time, it is also characterized by (v) the emergence of mechanistic and systems-level approaches to assess ENM safety, instead of just relying only on outdated chemical toxicology study methods and (vi) a growing awareness of the potential utility of alternative test strategies. These strategies decrease animal use and speed up the rate of discovery and knowledge generation by relying on more quantitative, mechanistic, and systems-level approaches that involve high throughput screening, computational modeling, and a variety of decision analysis approaches to improve regulatory decision-making[69].

While a variety of reasons explain why knowledge gathering and research cannot, as yet, address all the information requirements for hazard, exposure, dosimetry, and lifecycle analysis, there is a clear need to integrate and use available knowledge for safety assessment, decision analysis, and regulatory approval in light of the increasing number of products making their way to the marketplace. At the same time, the remaining knowledge gaps called out in the 2012 and 2014 NRC reports need to be addressed by well-planned and judicious research as well as obtaining information about the fate and lifecycle of nanomaterials entering the value-chain[66,67]. Focusing academia, industry, and the Federal Government together on leading industry sectors relying on nanotechnology is important to develop transparency and consensus on the nanotechnology EHS decision-making process that is required for safe implementation.

To continue reducing nanotechnology EHS uncertainty, expedite nanotechnology commercialization, and safe implementation of nanotechnology, the Environmental Protection Agency, National Institutes of Health, National Institute for Occupational Safety and Health, Food and Drug Administration, National Science Foundation, and National Institute of Standards and Technology should play a key coordinating role in (i) establishing multi-stakeholder Government, industrial, and academic communities that focus on nanotechnology EHS decision analysis in key commercial sectors dependent on ENM access to the marketplace and in (ii) integrated decision analysis, based on current available (but incomplete) data and information, while continuing to oversee the development and implementation of measurement tools for new and actionable data collection to improve the level of certainty and regulatory decision making.

In addition, to improve hazard and risk assessment, the nanotechnology community, together with appropriate regulatory authorities, should investigate how more rapid mechanistic approaches based on nanomaterial properties and interactions at the nanotechnology-biology interface can be used to replace chemical-oriented hazard assessment by tiered risk assessment approaches. Some of this testing should involve high throughput screening and computational analysis to speed up hazard ranking and establish predictive principles for toxicological analysis and safer design. The tiered risk assessment approaches should include both qualitative (e.g., listing of evidence, best professional judgment, weight of evidence, control banding, ENM categorization) and quantitative (e.g., occupational exposure limits, quantitative structure-activity relationships) measures to allow early as well as more advanced decision analysis, each of which has a role to play in deciding whether an ENM is a "material of concern" and/or how much additional testing is required to reach a decision on safe implementation[70].

To obtain more information about exposure in humans and the environment, the 2012 PCAST recommendation of "the importance of a life cycle approach to assessing risks" is still highly applicable, but its implementation by the lack of information about the volumes and types of materials in the commercial value chain. This outcome reflects, in part, the reluctance of industry to disclose information because of the uncertainty in regulatory decision-making and the fear of disclosing confidential business information. While France is attempting to solve this through a national nanomaterial registry, it is uncertain whether such an approach will succeed in the United States[71]. Thus, improving the ability to perform integrated decision analysis under the coordination of NNI becomes even more important to reduce uncertainty.

The Federal Government, with the assistance of industry, also needs to intensify research efforts to improve the instrumentation for assessing exposure and performing dosimetry analysis in humans and the natural environment (including potentially sensitive terrestrial, aquatic, estuarine and marine ecosystems, wastewater treatment systems, and end-of-life disposal sites). Lifecycle analysis and knowledge of materials in the value chain are also important for early and cost-effective decision analysis about the possible target sites to focus on. Activities such as those described previously presuppose a competent multi-disciplinary workforce. This workforce must continuously upgrade and improve its nanotechnology EHS expertise toward developing sustainable technologies and obtaining public trust and approval. The $121.1 million in nanotechnology EHS funding requested for FY2014 does not include funding for nanotechnology EHS test facilities where academia, industry, and Government can engage in EHS assessment and safer design of marketable products in the same way as the National Cancer Institute's Nanotechnology Cancer Laboratory is undertaking to foster cancer nanotherapeutics[72]. Some of the larger multi-disciplinary nanotechnology safety centers funded by NSF will sunset in 4 years, at which point no comprehensive, multidisciplinary communities will exist to continue the integrated research and education that is needed to continue to develop a U.S. nanotechnology EHS workforce.

Addressing the ethical, legal, and societal implications (ELSI) of nanotechnology in a proactive manner is critical to ensure public trust, especially in an era when the translation of critical research performed under NNI will lead to increased commercialization of nanotechnology[73]. NNI should continue to engage a variety of ELSI experts to gauge the societal impact, understanding, and perception of the added value of and concerns about nanotechnology. To increase stakeholder awareness and education concerning ELSI issues, appropriate and relevant ELSI information about the efforts to improve nanosafety should be disseminated. In addition to informing the public, the participation of members of the public in multi-stakeholder communities is critical.

Recommendation 12. NSET should continue to support the development of a multidisciplinary nanotechnology-EHS ecosystem that promotes non-animal based (alternative) test strategies for safety assessment and multi-stakeholder participation in regulatory decision-making and safe implementation to facilitate market access of nanomaterials and nanotechnology-enabled products by

a) Establishing multi-stakeholder working groups (e.g., semiconductors, carbon nanotubes and grapheme, silica) in which academia, industry, and Government experts use evidence-based nanotechnology EHS safety and risk analysis to facilitate the commercialization and entry of these materials and products to the marketplace;

b) Promoting the use of cost-effective rapid-throughput screening and mechanistic, predictive, and tiered risk assessment approaches that address the areas of greatest nanotechnology EHS concern and uncertainty;
c) Advocating for long-term and stable funding for nanotechnology EHS research, workforce development, knowledge acquisition, benign design, and nanotechnology EHS test facilities; and
d) Fostering the development of decision analysis methods and tools that make use of current knowledge for regulatory decision-making, while actively promoting actionable hazard, exposure, risk, and lifecycle analysis.

APPENDIX I. MEASURING THE CONTINUUM OF THE RESEARCH/TRANSLATION PROCESS AND SUBSEQUENT OUTCOMES USING STAR METRICS

Any attempt to create a data infrastructure around the effects of research and development (R&D) must confront the fact that relevant data (e.g., funding agency R&D awards, educational institution outcome data, research publications) are currently drawn from disparate sources, using widely differing methodologies and approaches. Thus, building a coherent data infrastructure is particularly challenging. Inputs, outputs, and outcomes are not currently generated or combined in a systematic fashion. The development of consistent and reliable answers to stakeholder requests requires the use of common data sources and standardized methodologies for data cleaning and analysis. Continuing to require research institutions and principal investigators to report research outcomes manually is neither practicable nor desirable. A recent study found that poorly integrated Federal reporting and other regulations impose a heavy and growing administrative burden on Federally funded research[74]. The report argues that this "regulatory overhead" is both large (and getting larger) and inefficient, with many Federal reporting requirements overlapping and even conflicting. It estimates that 42% of faculty time relating to Federally funded research is spent on administrative duties, rather than on the research itself. Not only is an approach using common data sources and consistent methodologies likely to reduce administrative burden, but the cost is likely to be negligible relative to other approaches. Experience with the Longitudinal Employer-Household Dynamics program showed that the cost of processing existing data was about one-half of a cent per record versus over $1,000 per record for survey data collection.

The project called Science and Technology for America's Reinvestment: Measuring the EffecT of Research on Innovation, Competitiveness, and Science (STAR METRICS) is one approach. The core of STAR METRICS is to use digital technologies to capture the data needed to understand and demonstrate the broad scientific, social, economic, and workforce results of Federal science and technology investments, as described in Figure A1. Research institutions are already developing structured information architectures to capture current and more accurate information about the interests, activities, and accomplishments of their scholars[75]. Science and technology agencies in other countries also have developed data platforms to identify and characterize scientific outcomes. An increasing volume and variety of research outputs, such as publications, patents, and datasets, are accessible in digital form and are harvested via services such as Citeseer, Google Scholar, and Microsoft Academic Search. Increasingly accurate methods exist for reliably attributing research products to researchers, a nontrivial task due to considerable ambiguity in author names[76,77].

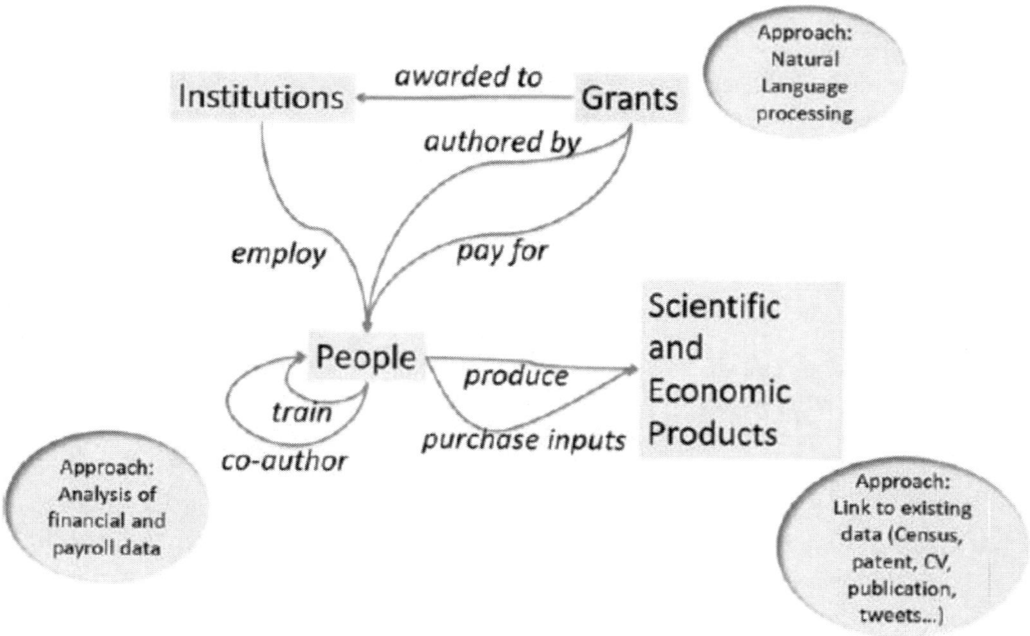

Figure A1. A schematic approach to the analysis of the results of science funding[78].

There are many ways to apply new technologies (Figure A1). The first example is to use tools like natural language processing, rather than manual classification, to describe *what* research is being done using automated analysis of proposal and award text to identify the research topics in a portfolio[79]. The second is to use administrative records to describe *who* is doing research, *with whom*, and with what inputs[39]. Thus, for example, STAR METRICS data from eight Committee on Institutional Cooperation (CIC) institutions have been used to generate a report documenting the research workforce and the short-run economic impacts of Federal research at CIC institutions[39]. The third is to use curricula vitae and other sources of data to describe *what results* the funding has generated[42,80-82]. One example is the prototype R&D Dashboard (http://rd-dashboard.nitrd.gov/), which was supported by the White House Chief Technology Office and the Chief Information Office.

In practice, this approach means that many measures of scientific and economic activity can be directly generated from existing data, rather than hand curated. For example, in an in-depth and detailed study of the Center for Environmentally Responsible Solvents and Processes at the University of North Carolina, a full time staff person who kept track of statistical information collected much of the data[83]. But many of these direct measures can be automatically generated using a fully-fledged STAR METRICS platform.

APPENDIX II. MANUFACTURING NANOMEDICINE

Fundamental research has advanced nanotechnology in sectors that will require investing in manufacturing infrastructure to successfully commercialize these technologies. Since much of biology carries out at the nanoscale or is mediated by nanoscale structures, the advent of nanotechnology opens up the possibility of bringing a fundamental, engineered approach to medicine. The field of "nanomedicine" has the potential to produce biologically interactive nanoscale materials that prevent, diagnosis, and treat disease; maintain and improve human health; and regenerate tissue. Nano-based approaches have significant advantages over traditional approaches in their potential to modulate biological systems.

A broad array of novel technologies are being developed that have potential to transform the diagnosis and treatment of human disease:

- Nanofabrication technologies can now efficiently generate nanostructures with defined shapes and sizes that can influence particle distribution through the body and uptake by different cell types[84]. This provides a promising technical foundation for developing particle-based approaches for delivering drug payloads to different diseased tissues throughout the body.
- There has been continuous advancement and refinement of nanocarriers that deliver cancer chemotherapeutic agents, including short nucleic-acid oligomers anchored to gold nanoparticles that are taken up quickly and safely into a variety of cell types; multifunctional nanoparticles consisting of self-assembling polymers or micelles that can image and/or deliver synergistic drug combinations; and on-demand release particles that can be autonomously (e.g., pH, temperature) or remotely (e.g., magnetically) controlled to release drug payloads[85-89].
- The biggest advance in diagnostics has been the development of highly sensitive nano- enabled lab-on-chip devices that can detect biomarkers of disease at the femtomol and attomol detection levels and multiplexing portable devices that can give instantaneous readout of biomarkers of disease at the bedside or in the home.
- Nanotechnology is being widely used for antibacterial applications in the hospital environment and is also poised to affect the treatment of infectious disease, e.g., by physically damaging bacterial cell walls and biofilms with mechanisms that bypass antibiotic resistance[90, 91].
- Nanomaterials can have unique capabilities to provide immune modulation, antigen delivery, and immunostimulatory effects[92,93]. Nanotechnology is also influencing vaccine development for infectious diseases and cancer.
- Novel coatings can be applied to the nanoparticle surfaces to facilitate rapid particle movement across natural, protective mucous barriers. This strategy offers the potential of highly effective delivery approaches to topically treat eye, lung, and gut diseases[94].

Although these examples of nano-based therapies may be fundamentally different in structure and function, they have to overcome common challenges during review and registration by the Food and Drug Administration – the development of defined manufacturing methods that can achieve regulatory approval. A significant emphasis must be placed early in the commercialization pathway on refining or replacing laboratory fabrication procedures with reliable, consistent, and economically viable manufacturing methods that can be scaled up for clinical development and, ultimately, to reliably generate commercial drug supply. Start-ups frequently must focus considerable time and capital on developing these methods. The need to scale up is likely to occur early on the product development timeline since animal testing in the relevant disease models and understanding of how these nanotherapies are distributed in the body are required for making the decision to proceed toward clinical development.

Another important consideration in manufacturing nanomedicines is the need to conduct preclinical toxicity studies and any subsequent clinical trials using drug supply generated under Good Manufacturing Practice (GMP) conditions in an approved facility. Thus this investment must be made prior to knowing whether the nanomedicine will be effective in humans for its intended indication. Depending on the novelty and complexity of the manufacturing process, there may be few options to source the manufacturing to outside parties. In the absence of established nanomanufacturing facilities in which to explore methods, for preclinical studies, complete method validation, or deploy a GMP-based manufacturing scale-up protocol, the start-up has no alternative but do this internally. Clearly, access to manufacturing facilities for scaling up nanomedicines in the amounts necessary for animal testing and preclinical development would accelerate the transition of these novel therapies to proof-of-concept human testing in clinical development.

APPENDIX III. REVIEW PARTICIPANTS

PCAST wishes to express gratitude to the following individuals who contributed to this assessment.

Eric Amis
Director, Physical Sciences
United Technologies Research Center

Thomas Bartholomew
Program Examiner
Office of Management and Budget

Adam Bobrow
Senior Policy Advisor for International Affairs
Office of Science and Technology Policy

Chris Cannizzaro
Foreign Affairs Officer Department of State

Altaf Carim
Assistant Director, Nanotechnology Office of Science and Technology Policy

Charina Choi
White House Fellow
Office of Science and Technology Policy

David Danielson
Assistant Secretary for Energy Efficiency and Renewable Energy Department of Energy

Krishna Doraiswamy
Research Collaborations Manager DuPont Science and Technology

Cristin Dorgelo Chief of Staff
Office of Science and Technology Policy

Michael Fancher
Vice President, Business Development and Economic Outreach
SUNY College of Nanoscale Science and Engineering

Piotr Grodzinski
Director
Office of Cancer Nanotechnology Research

Ron Jarmin
Assistant Director for Research and Methodology
U.S. Census Bureau

Lloyd Johnston
Vice President, Pharmaceutical Research and Development
Selecta Biosciences, Inc.

Tom Kalil
Deputy Director for Technology and Innovation
Office of Science and Technology Policy

Pramod Khargonekar
Assistant Director, Engineering National Science Foundation

Fred Klaessig Manager
Pennsylvania Bio Nano Systems

Harriet Kung
Associate Director of Science for Basic Energy Sciences
U.S. Department of Energy

Lerwen Liu
Nanotechnology Expert
Asia Technology Information Program

Doug Lowy
Deputy Director
National Cancer Institute

George Maracas
Program Manager, Nanoscale Science Research Centers
Department of Energy

Scott McNeil
Director
Nanotechnology Characterization Laboratory

Don Millard
Program Director
National Science Foundation

JoAnn Milliken
Hydrogen Program Manager Department of Energy

Jayne Morrow
NSTC Executive Director
Office of Science and Technology Policy

Sarah Mueller
Graduate Research Assistant
University of North Carolina, Chapel Hill

Venkatesh Narayanamurti
Director
Harvard Kennedy School Program in Science,
Technology and Public Policy

Mihail Roco
Senior Advisor for Nanotechnology National Science Foundation

Neil Shinn
Director, Center for Integrated Nanotechnologies
Los Alamos National Laboratory

Robin Staffin
Director for Basic Research U.S. Department of Defense

Tom Theis
Program Manager
New Devices and Architectures for Computing
IBM

Sally Tinkle
Senior Strategy and Policy Manager Science & Technology Policy Institute

Lloyd Whitman
Interim Director National Nanotechnology Coordination Office

Jan Youtie
Director, Georgia Tech Program in Science, Technology, and Innovation Policy
Georgia Institute of Technology

End Notes

[1] Llordes, A., et al., Tunable near-infrared and visible-light transmittance in *nanocrystal-in-glass composites.* Nature, 2013. 500(7462): p. 323-326.

[2] Bailey, M., et al., *Functionalized, carbon nanotube material for the catalytic degradation of organophosphate nerve agents.* Nano Research, 2014. 7(3): p. 390-398.

[3] 108th Congress, *21st Century Nanotechnology Research and Development Act* in *Public Law 108- 153*. 2003.

[4] President's Council of Advisors on Science and Technology, *Report to the President and Congress on the Fourth Assessment of the National Nanotechnology Initiative* 2012.

[5] United States Government Accountability Office, *Nanotechnology - Improved Performance Information Needed for Environmental, Health, and Safety Research*. 2012.

[6] National Research Council, *Triennial Review of the National Nanotechnology Initiative*. 2013.

[7] United States Government Accountability Office, *Nanomanufacturing: Emergence and Implications for U.S. Competitiveness, the Environment, and Human Health*. 2014.

[8] Roco, M.C., C.A. Mirkin, and M.C. Hersam, *Nanotechnology Research Directions for Societal Needs in 2020*. 2010, World Technology Evaluation Center.

[9] The NSET is a subcommittee of the Committee on Technology of the National Science and Technology Council (NSTC), which is responsible for coordinating science and technology policy across the Federal Government.

[10] *National Nanotechnology Initiative - Frequently Asked Questions*. Available from: http://www.nano.gov /nanotech-101/nanotechnology-facts.

[11] "Chaired by the President, the membership of the NSTC is made up of the Vice President, the Director of the Office of Science and Technology Policy, Cabinet Secretaries and Agency Heads with significant science and technology responsibilities, and other White House officials." http://www.whitehouse.gov /administration/eop /ostp/nstc

[12] Current representation on NSET includes the Consumer Product Safety Commission; the Bureau of Industry and Security, the Economic Development Administration, the National Institute of Standards and Technology, and the U.S. Patent and Trademark Office, all from the Department of Commerce; the Department of Defense; the Department of Education; the Department of Energy; the Agency for Toxic Substances and Disease Registry, the Food and Drug Administration, the National Institute for Occupational Safety and Health, and the National Institutes of Health, all from the Department of Health and Human Services; the Department of Homeland Security; the U.S. Geological Survey from the Department of the Interior; the National Institutes of Justice from the Department of Justice; the Occupational Safety and Health Administration from the Department of Labor; the Department of State; the Department of Transportation, the Department of the Treasury; the Environmental Protection Agency; the Office of the Director of National Intelligence and the National Reconnaissance Office; the National Aeronautics and Space Administration; the National Science Foundation; the Nuclear Regulatory Commission; the Agriculture Research Service, the Forest Service, and the National Institute of Food and Agriculture from the U.S. Department of Agriculture; and the U.S. International Trade Commission. http://www.nano.gov/nset

[13] *National Nanotechnology Initiative - Working Groups & Coordinators* Available from: http://www.nano.gov /about-nni/working-groups.

[14] *National Nanotechnology Initiative - National Nanotechnology Coordination Office*. Available from: http://www.nano.gov/about-nni/nnco.

[15] *National Nanotechnology Initiative - NNI Organizational Chart*. Available from: http://nano.gov/node/1115.

[16] Subcommittee on Nanoscale Science, Engineering, and Technology; Committee on Technology; National Science and Technology Council, *National Nanotechnology Initiative: Supplement to the President's 2015 Budget*. 2014.

[17] PCAs are a legacy mechanism for understanding how NNI funding is allocated.

[18] National Research Council, *Interim Report for the Triennial Review of the National Nanotechnology Initiative, Phase II*. 2013.

[19] Roco, M., *The Long View of Nanotechnology Development: the National Nanotechnology Initiative at 10 years*. Journal of Nanoparticle Research, 2011. 13: p. 427-445.

[20] Nieto, M., F. Lopez, and F. Cruz, *Performance Analysis of Technology Using the S Curve Model: the Case of Digital Signal Processing (DSP) Technologies*. Technovation, 1998. 18(6-7): p. 439- 457.

[21] Kisliuk, B., J. Lorengo, and G. Corcoran, *Individual communication*. 2013: United States Patent and Trademark Office.

[22] Flynn, H., *Nanotechnology Update: Corporations Up Their Spending as Revenues for Nano-enabled Products Increase*. 2014, Lux Research.

[23] Data represents best estimates from funding agency websites, center websites, and direct conversations with program manager or university center contacts.

[24] Subcommittee on Nanoscale Science, Engineering, and Technology; Committee on Technology; National Science and Technology Council, *National Nanotechnology Initiative: Strategy for Nanotechnology-Related Environmental, Health, and Safety Research*. 2008.

[25] In his NNI-launching speech, President Clinton remarked, "Some of these [nanotechnology] research goals will take 20 or more years to achieve. But that is why... there is such a critical role for the Federal Government. It may take decades."

[26] Source: V. Li, S. Arora, J. Voutie, P. Shapira, S. Carley, Program on Nanotechnology Research and Innovation System Assessment, Georgia Institute of Technology, Atlanta, Georgia. Analysis of publication records of articles indexed in ThomsonReuters Web of Science (SCI, SSCI, and A&HI), March 2014. This material is based upon work supported by the National Science Foundation (NSF) through the Center for Nanotechnology in Society (CNS-ASU) under NSF Grant Number 0937591.

[27] Arora, S., et al., *Capturing new developments in an emerging technology: an updated search strategy for identifying nanotechnology research outputs*. Scientometrics, 2013. 95(1): p. 351- 370.

[28] Subcommittee on Nanoscale Science, Engineering, and Technology; Committee on Technology; National Science and Technology Council, *National Nanotechnology Initiative: Strategic Plan*. 2014.

[29] Three NSIs were established in 2011: Sustainable Nanomanufacturing (Nanomanufacturing), Solar Energy Collection and Conversion (Solar), and Nanoelectronics for 2020 and Beyond (Nanoelectronics). In 2012, two more were initiated: Nanotechnology for Sensors and Sensors for Nanotechnology (Sensors), and Nanotechnology Knowledge Infrastructure (NKI). http://nano.gov/signatureinitiatives

[30] NSI White Papers. Available from: http://nano.gov/signatureinitiatives

[31] More specifically, Nanoelectronics and NKI address all characteristics at least partially, while Solar, Nanomanufacturing, and Sensors do not address resources, investments, risk management, integration, or implementation. The Solar white paper does not contain a plan for interagency collaboration, and the Committee found little evidence that interaction was occurring. The Committee determined that Nanomanufacturing and Nanoelectronics were effectively collaborating and communicating with Federal and non-Federal partners, defining metrics of success with timelines and outcomes, and conducting periodic progress reviews and adjusting their plans accordingly.

[32] Department of Energy, *DOE Pursues SunShot Initiative to Achieve Cost Competitive Solar Energy by 2020*. 2011.

[33] Although bibliometrics have come to dominate science metrics, they do not measure scientific knowledge.

[34] Lane, J., *Let's Make Science Metrics More Scientific*. Nature, 2010. 464: p. 488-489.

[35] Lane, J., M. Largent, and R. Rosen, *Science Metrics and Science Policy*, in *Next Generation Metrics: Harnessing Multidimensional Indicators of Scholarly Performance*, B. Cronin and C. Sugimoto, Editors. 2013, MIT Press.

[36] Jaffe, A.B. and J. Lerner, *Innovation and Its Discontents: How Our Broken Patent System is Endangering Innovation and Progress, and What to Do About It*. Vol. 2nd edition. 2007, Princeton, NJ: Princeton University Press.

[37] As has been pointed out by many others, comparing the results of research expenditures with a counterfactual of what would have been the results without the expenditures, which is easier said than done, is critical.

[38] Science and Technology for America's Reinvestment—Measuring the Effects of Research on Innovation, Competitiveness, and Science. The STAR METRICS partnership initially emerged as a result of reporting requirements associated with the 2009 American Recovery and Reinvestment Act; it developed in response to a

recognized need to begin to systematically document Federal investments in science and their immediate and long-term results with the analytical focus being on the activities of scientists rather than on counting publications.

[39] Weinberg, B.A., et al., *Science Funding and Short-Term Economic Activity.* Science, 2014. 344(6179): p. 41-43.

[40] Nelson, L. and S.W. Sedwick, *STAR METRICS: A Participant's Perspective*, in *NCURA Magazine.* 2011, National Council of University Research Administrators.

[41] The program, inspired by the Longitudinal Employer-Household Dynamics program at the Census Bureau, developed an automated approach to capturing project level payroll data on all individuals paid by Federal science funding at each participating research organization. Data were also collected on all payments to vendors and subcontracts, as well as overhead expenditures. 14 data elements were collected. Details are available at https://starmetrics.nih.gov.

[42] Evans, J.A. and J.G. Foster, *Metaknowledge.* Science, 2011. 331: p. 721-725.

[43] While some start-ups successfully bring products to the marketplace, in many cases they license their early-stage product prototypes to a larger technology or pharmaceutical company, or small companies with exciting technology platforms may simply be acquired by larger corporations. Instead of an independent entrepreneur spinning off a start-up company, a large technology and pharmaceutical company could directly fund an academic investigator working in areas of mutual interest and commercialize the result itself.

[44] Huang, K.G. and F.E. Murray, *Entrepreneurial Experiments in Science Policy: Analyzing the Human Genome Project* Research Policy, 2010. 39(5): p. 567-582.

[45] Jensen, R., J. Thursby, and M.C. Thursby, *University-Industry Spillovers, Government Funding, and Industrial Consulting.* National Bureau of Economic Research Working Paper Series, 2010. No. 15732.

[46] Murray, F., *Innovation as Co-evolution of Scientific and Technological Networks: Exploring Tissue Engineering.* Research Policy, 2002. 31(8–9): p. 1389-1403.

[47] Thursby, J., A.W. Fuller, and M. Thursby, *US Faculty Patenting: Inside and Outside the University.* Research Policy, 2009. 38(1): p. 14-25.

[48] Thursby, J.G. and M.C. Thursby, *Who Is Selling the Ivory Tower? Sources of Growth in University Licensing.* Management Science, 2002. 48(1): p. 90-104.

[49] SRC: Semiconductor Research Corporation. Available from: https://www.src.org/.

[50] Berznitz, D. and M. Murphree, *Run of the Red Queen: Government, Innovation, Globalization, and Economic Growth in China.* 2011: Yale University Press.

[51] Fuchs, E. and R. Kirchain, *Design for Location? The Impact of Manufacturing Offshore on Technology Competitiveness in the Optoelectronics Industry.* Management Science, 2010. 56(12): p. 2323-2349.

[52] Fuchs, E.R.H., et al., *Plastic Cars in China? The Significance of Production Location Over Markets for Technology Competitiveness in the United States Versus the People's Republic of China.* International Journal of Production Economics, 2011. 132(1): p. 79-92.

[53] "In a Manufacturing Innovation Institute (MII) hub, industry, academia, and government partners leverage existing resources, collaborate, and co-invest to nurture manufacturing innovation and accelerate commercialization network is proposed to commercialization. As sustainable manufacturing innovation hubs, MIIs will create, showcase, and deploy new capabilities, new products, and new processes that can impact commercial production....The program addresses the reality that the substantial Federal investment in basic research isn't enough to make sure that a new technology crosses the bridge all the way from invention to product development and process prototyping to manufacturing at scale and commercializing, the stages of wealth and job creation." http://manufacturing.gov/nnmi.html

[54] (1) The National Additive Manufacturing Innovation Institute (NAMII), now known as America Makes, headquartered in Youngstown, Ohio; (2) the Next Generation Power Electronics Manufacturing Innovation Institute at North Carolina State University in Raleigh, North Carolina; (3) the Digital Manufacturing and Design Innovation Institute in Chicago, Illinois led by UI Labs; and (4) the Lightweight and Modern Metals Manufacturing Innovation Institute in Detroit, Michigan, led by EWI.

[55] National Science Foundation: Innovation Corps. Available from: http://www.nsf.gov/news/specialreports/i-corps/.

[56] President's Council of Advisors on Science and Technology, *Report to the President, Transformation and Opportunity: The Future of the U.S. Research Enterprise.* 2012.

[57] Department of Commerce, *The Innovative and Enterpreneurial University: Higher Education, Innovation & Entrepreneurship in Focus.* 2013.

[58] Council On Governmental Relations, *A Tutorial on Technology Transfer in U.S. Colleges and Universities.* 2011.

[59] Feldman, M.P., A. Colaianni, and C. Kang Liu, *Lessons From the Commercialization of the Cohen-Boyer Patents: The Stanford University Licensing Program*, in *Intellectual property management in health and*

agricultural innovation: a handbook of best practices, A. Krattiger, et al., Editors. 2007, MIHR-USA p. 1797–1807.

[60] National Research Council, *Managing University Intellectual Property in the Public Interest*. 2010.

[61] Subcommittee on Nanoscale Science, Engineering, and Technology; Committee on Technology; National Science and Technology Council. *Regional, State, and Local Initiatives in Nanotechnology: Report of the National Nanotechnology Initiative Workshop*. 2009. Oklahoma City, OK.

[62] Subcommittee on Nanoscale Science, Engineering, and Technology; Committee on Technology; National Science and Technology Council. *Regional, State, and Local Initiatives in Nanotechnology: Report of the National Nanotechnology Initiative Workshop*. 2012. Portland, OR.

[63] Tremblay, J.-F., *American Takes Charge In Tianjin*, in *Chemical & Engineering News*. 2013, American Chemical Society.

[64] Skaarup, G., *World Leading Researcher and New Research Center to the Niels Bohr Institute*. 2011, University of Copenhagen.

[65] National Nanotechnology Infrastructure Network, *Annual Report of the National Nanotechnology Infrastructure Network (abridged)*. 2005.

[66] National Research Council, *A Research Strategy for Environmental, Health, and Safety Aspects of Engineered Nanomaterials*. 2012.

[67] National Research Council, *Research Progress on Environmental, Health, and Safety Aspects of Engineered Nanomaterials*. 2014.

[68] Subcommittee on Nanoscale Science, Engineering, and Technology; Committee on Technology; National Science and Technology Council, *Progress Review on the Coordinated Implementation of the National Nanotechnology Initiative 2011 Environmental, Health, and Safety Research Strategy*. 2014.

[69] Collins, F.S., G.M. Gray, and J.R. Bucher, *Toxicology. Transforming Environmental Health Protection*. Science, 2008. 319(5865): p. 906-907.

[70] Nel, A.E., et al., *A Multi-Stakeholder Perspective on the Use of Alternative Test Strategies for Nanomaterial Safety Assessment*. ACS Nano, 2013. 7(8): p. 6422-6433.

[71] French Ministry of Ecology, Sustainable Development, Transport, and Housing, *On the Annual Declaration of Substances at Nanoscale in Application of Article R. 523-4 of the Environment Code*, in *Decree no. 2012-232*. 2012: Official Journal of the French Republic.

[72] *National Cancer Institute - Nanotechnology Characterization Laboratory*. Available from: http://ncl.cancer.gov/.

[73] Roco, M.C., et al., *Innovative and Responsible Governance of Nanotechnology for Societal Development*, in *Nanotechnology Research Directions for Societal Needs in 2020*. 2011, Springer: Boston.

[74] Smith, T.L., et al., *Reforming Regulation of Research Universities*. Issues in Science and Technology, 2012: p. 57-64.

[75] Including, for example, the VIVO Project http://vivoweb.org, the Harvard Profiles System, and others.

[76] Han, H., et al., *Two Supervised Learning Approaches for Name Disambiguation in Author Citations*. Proceedings of the 2004 Joint ACM/IEEE Conference on Digital Libraries, 2004: p. 296- 305.

[77] Smalheiser, N.R. and V. Torvik, *Author Name Disambiguation*. Annual Review of Information Science and Technology, 2009: p. 1-43.

[78] Foster, Ian and Julia Lane, Science Based Measures of Science Investments, working paper.

[79] Edmund, D.N., et al., *Database of NIH Grants Using Machine-Learning Categories and Graphical Clustering*. Nature Methods, 2011. 8: p. 443-444.

[80] Evans, J.A., *Future Science*. Science, 2013. 342: p. 44-45.

[81] Giles, C.L., K.D. Bollacker, and S. Lawrence, *CiteSeer: An Automatic Citation Indexing System*. Proceedings of the Third ACM Conference on Digital Libraries, 1998: p. 89-98.

[82] Khabsa, M., P. Treeratpitu, and L. Giles, *AckSeer: A Repository and Search Engine for Automatically Extracted Acknowledgments from Digital Libraries*. Proc. 2004 Joint ACM/IEEE Conference on Digital Libraries, 2012: p. 89-98.

[83] Feldman, M., A. Freyer, and L. Lanahan, *On the Measurement of University Research Contribution to Economic Growth and Innovation*, in *Universities and Colleges as Economic Drivers*, J. Lane and B. Johnstone, Editors. 2012, SUNY series, Critical Issues in Higher Education.

[84] *Liquidia Technologies*. Available from: http://www.liquidia.com/ProductPlatform. html.

[85] *AuraSense Therapeutics*. Available from: http://www.aurasensetherapeutics.com/.

[86] Choi, C.H.J., et al., *Mechanism for the Endocytosis of Spherical Nucleic Acid Nanoparticle Conjugates*. Proceedings of the National Academy of Sciences, 2013. 110(19): p. 7625-7630.

[87] Zeng, D., et al., *Topical Delivery of siRNA-based Spherical Nucleic Acid Nanoparticle Conjugates for Gene*

Regulation. Proceedings of the National Academy of Sciences, 2012. 109(30): p. 11975- 11980.

[88] Meng, H., et al., *Engineered Design of Mesoporous Solica Nanoparticles to Deliver Doxorubicin and P-Glycoprotein stRNA to Overcome Drug Resistance in a Cancer Cell Line.* Journal of the American Chemical Society, 2010. 4(8): p. 4539-4550.

[89] Thomas, C.R., et al., *Noninvasive Remote-Controlled Release of Drug Molecules in Vitro Using Magnetic Actuation of Mechanized Nanoparticles.* Journal of the American Chemical Society, 2010. 132(31): p. 10623-10625.

[90] Liong, M., et al., *Antimicrobial Activity of Silver Nanocrystals Encapsulated in Mesoporous Silica Nanoparticles.* Advanced Materials, 2009. 21(17): p. 1684-1689.

[91] Sanders, R., *Discovery Opens Door to Attacking Biofilms That Cause Chronic Infections*. 2012, UC Berkeley News Center.

[92] Sun, B., et al., *Engineering an Effective Immune Adjuvant by Designed Control of Shape and Crystallinity of Aluminum Oxyhydroxide Nanoparticles.* ACS Nano, 2013. 7(12): p. 10834-10849.

[93] Ulery, B.D., et al., *Rational Design of Pathogen-Mimicking Amphiphilic Materials as Nanoadjuvants.* Scientific Reports, 2011. 1(198).

[94] Wang, Y.-Y., et al., *Addressing the PEG Mucoadhesivity Paradox to Engineer Nanoparticles that "Slip" through the Human Mucus Barrier.* Angewandte Chemie International Edition, 2008. 47: p. 9726-9729.

In: The National Nanotechnology Initiative
Editor: Marlin Thorne

ISBN: 978-1-63463-694-0

Chapter 2

THE NATIONAL NANOTECHNOLOGY INITIATIVE: SUPPLEMENT TO THE PRESIDENT'S 2015 BUDGET*

National Science and Technology Council,
Committee on Technology,
Subcommittee on Nanoscale Science,
Engineering, and Technology

About the National Science and Technology Council

The National Science and Technology Council (NSTC) is the principal means by which the Executive Branch coordinates science and technology policy across the diverse entities that make up the Federal research and development enterprise. A primary objective of the NSTC is establishing clear national goals for Federal science and technology investments. The NSTC prepares research and development strategies that are coordinated across Federal agencies to form investment packages aimed at accomplishing multiple national goals. The work of the NSTC is organized under committees that oversee subcommittees and working groups focused on different aspects of science and technology. More information is available at www.WhiteHouse.gov/administration/eop/ostp/nstc.

About the Office of Science and Technology Policy

The Office of Science and Technology Policy (OSTP) was established by the National Science and Technology Policy, Organization, and Priorities Act of 1976. OSTP's responsibilities include advising the President in policy formulation and budget development on questions in which science and technology are important elements; articulating the President's science and technology policy and programs; and fostering strong partnerships among Federal, state, and local governments, and the scientific communities in industry and academia. The Director of OSTP also serves as Assistant to the President for Science and Technology and manages the NSTC. More information is available at www.ostp.gov.

---

* This is an edited, reformatted and augmented version of a report issued March 2014.

About this document

This document is a supplement to the President's 2015 Budget request submitted to Congress on March 4, 2014. It gives a description of the activities underway in 2013 and 2014 and planned for 2015 by the Federal Government agencies participating in the National Nanotechnology Initiative (NNI), primarily from a programmatic and budgetary perspective. It is based on the NNI Strategic Plan released in February 2014 and reports actual investments for 2013, estimated investments for 2014, and requested investments for 2015 by Program Component Area (PCA), as called for under the provisions of the 21st Century Nanotechnology Research and Development Act of 2003 (Public Law 108-153, 15 USC §7501). The report also addresses the requirement for Department of Defense reporting on its nanotechnology investments, per 10 USC §2358. Additional information regarding the NNI is available on the NNI website at www.nano.gov.

About the cover

The cover images depict silica-core, gold-shell nanoparticles (gold nanoshells) as visualized by a transmission electron microscope (TEM). Each particle consists of a silica core (on the order of 100 nanometers in diameter) surrounded by a gold shell that is approximately 15 nanometers thick. These particles have unique features because of their nanoscale size and composition, and emit heat when activated with a specific wavelength of light. These gold nanoshells are currently undergoing clinical trial for targeted thermal ablation of tumors in cancer therapy. Preclinical characterization of these nanoparticles was performed at the Nanotechnology Characterization Laboratory, in collaboration with Nanospectra Biosciences, Inc., at the Frederick National Laboratory for Cancer Research, funded by the National Cancer Institute. The National Science Foundation and the Department of Defense provided early institutional support to Rice University and Nanospectra for research on the development of the gold nanoshells and exploration of their potential applications.

Front: A "slice" from a 3D tomography image of the gold nanoshells, with false color added to emphasize the gold shell (yellow) and silica core (blue). To create this 3D reconstruction, the sample grid was gradually tilted, acquiring TEM images at every 2°. The resulting series of images were correlated and reconstructed into a 3D model. This technique allows researchers to visualize the inside of the particles and to examine the completeness of the gold shell.

Back: TEM of the same gold nanoshells without 3D rendering. Cover and book design

Cover design is by Kathy Tresnak of Koncept, Inc. Book design is by staff of the National Nanotechnology Coordination Office (NNCO).

Copyright information

Printed in the United States of America, 2014.

EXECUTIVE OFFICE OF THE PRESIDENT
NATIONAL SCIENCE AND TECHNOLOGY COUNCIL
WASHINGTON, D.C. 20502
March 24, 2014

Dear Members of Congress:

I am pleased to forward this annual report on the multi-agency National Nanotechnology Initiative (NNI), in the form of the NNI Supplement to the President's Budget for Fiscal Year 2015. This document summarizes the programs and coordinated activities taking place across the many departments, agencies, and commissions participating today in the NNI – an initiative that has been a leading model of Federal science and technology coordination for over thirteen years.

The proposed NNI budget for Fiscal Year 2015 of over $1.5 billion will continue to advance our understanding of nanoscale phenomena and our ability to engineer nanoscale devices and systems that address national priorities and global challenges. As is true across the spectrum of Federal science and technology investments, the FY 2015 NNI budget reflects difficult choices as well as the maturation and evolution of programs at some agencies. Despite current fiscal constraints, robust efforts are maintained in key investment areas such as the nanotechnology signature-initiative areas and environmental, health, and safety research. NNI activities also intersect with and support national priorities that are not confined to the domain of nanotechnology, including advanced manufacturing, the Materials Genome Initiative (a drive to accelerate the design and deployment of advanced materials), and the BRAIN (Brain Research through Advancing Innovative Neurotechnologies) Initiative.

The NNI investment sustains vital support for fundamental, ground-breaking R&D, research infrastructure (including world-class centers, networks, and user facilities), and education and training programs that collectively constitute a major U.S. innovation enterprise. Signature initiatives within the NNI leverage programmatic collaborations among multiple agencies to place particular emphasis on jointly identified opportunities. These signature initiatives include ongoing activities in solar energy, next-generation electronics, sustainable manufacturing, nanoinformatics and modeling, and sensors. Pursuant to a revision of the NNI's Program Component Areas, detailed information on agency investments in signature initiative areas is now included in the annual budget reporting.

It is essential that the United States continue to lead the way in innovation enabled by nanotechnology and other emerging technologies. Now more than ever, the Nation's economic growth and global competitiveness depend on it. Thank you for sharing and supporting that vision.

Sincerely,

John P. Holdren
Director and
Assistant to the President for Science and Technology

1. Introduction and Overview

Overview of the National Nanotechnology Initiative

The National Nanotechnology Initiative (NNI), established in 2001,[1] is a U.S. Government research and development (R&D) initiative involving 20 Federal departments and independent agencies and commissions[2] working together toward the shared and challenging vision of "*a future in which the ability to understand and control matter at the nanoscale leads to a revolution in technology and industry that benefits society.*"[3] The combined, coordinated efforts of these agencies have accelerated discovery, development, and deployment of nanotechnology to benefit agency missions in service of the broader national interest. The 20 Federal agencies participating in the NNI are shown in Table 1; 11 of these report specific budget data for nanotechnology R&D for 2015.

The NNI is managed within the framework of the National Science and Technology Council (NSTC), the Cabinet-level council by which the President coordinates science and technology policy across the Federal Government. The Nanoscale Science, Engineering, and Technology (NSET) Subcommittee of the NSTC's Committee on Technology coordinates planning, budgeting, program implementation, and review of progress for the Initiative. The NSET Subcommittee is composed of representatives from participating agencies and the Executive Office of the President. A listing of official NSET Subcommittee members is provided at the front of this report, and contact information for NSET Subcommittee participants is provided in Appendix C. The National Nanotechnology Coordination Office (NNCO) acts as the primary point of contact for information on the NNI; provides technical and administrative support to the NSET Subcommittee, including the preparation of multiagency planning, budget, and assessment documents; develops, updates, and maintains the NNI website www.nano.gov; and provides public outreach on behalf of the NNI.

What is Nanotechnology?

Nanotechnology is the understanding and control of matter at dimensions between approximately 1 and 100 nanometers, where unique phenomena enable novel applications. Encompassing nanoscale science, engineering, and technology, nanotechnology involves imaging, measuring, modeling, and manipulating matter at this length scale.

A nanometer is one-billionth of a meter. A sheet of paper is about 100,000 nanometers thick; a single gold atom is about a third of a nanometer in diameter. Dimensions between approximately 1 and 100 nanometers are known as the nanoscale. Unusual physical, chemical, and biological properties can emerge in materials at the nanoscale. These properties may differ in important ways from the properties of bulk materials and single atoms or molecules.

Table 1. Federal Departments and Agencies Participating in the NNI

Federal departments and independent agencies and commissions with budgets dedicated to nanotechnology research and development (11)
Consumer Product Safety Commission (CPSC)†
Department of Commerce (DOC)
National Institute of Standards and Technology (NIST)
Department of Defense (DOD)
Department of Energy (DOE)
Department of Health and Human Services (DHHS)
Food and Drug Administration (FDA)
National Institute for Occupational Safety and Health (NIOSH)
National Institutes of Health (NIH)
Department of Homeland Security (DHS)
Department of Transportation (DOT)
Federal Highway Administration (FHWA)
Environmental Protection Agency (EPA)
National Aeronautics and Space Administration (NASA)
National Science Foundation (NSF)
U.S. Department of Agriculture (USDA)
Agricultural Research Service (ARS)
Forest Service (FS)
National Institute of Food and Agriculture (NIFA)
Other participating departments and independent agencies and commissions (9)
Department of Education (DOEd)
Department of the Interior (DOI)
U.S. Geological Survey (USGS)
Department of Justice (DOJ)
National Institute of Justice (NIJ)
Department of Labor (DOL)
Occupational Safety and Health Administration (OSHA)
Department of State (DOS)
Department of the Treasury (DOTreas)
Intelligence Community (IC)
Office of the Director of National Intelligence (ODNI)
National Reconnaissance Office (NRO)
Nuclear Regulatory Commission (NRC)†
U.S. International Trade Commission (USITC)†
Also participating from the Department of Commerce (DOC), listed above:
Bureau of Industry and Security (BIS)
Economic Development Administration (EDA)
U.S. Patent and Trademark Office (USPTO)

KEY † Denotes an independent commission that is represented on NSET but is non-voting.

The NSET Subcommittee has identified some critical issues that require special attention to interagency coordination, facilitated through subsidiary working groups and/or individual coordinator functions. NSET has therefore established two working groups—the

Nanotechnology Environmental and Health Implications (NEHI) Working Group, and the Nanomanufacturing, Industry Liaison, and Innovation (NILI) Working Group—and four coordinators—for global issues; standards development; environmental, health, and safety research; and education, engagement, and societal dimensions. For additional details on these working groups and coordinators, see the Coordination and Assessment section of the 2014 NNI Strategic Plan.[4]

The 2014 NNI Strategic Plan sets out the vision for the NNI stated above and specifies four goals aimed at achieving that overall vision:

1) Advance a world-class nanotechnology research and development program.
2) Foster the transfer of new technologies into products for commercial and public benefit.
3) Develop and sustain educational resources, a skilled workforce, and a dynamic infrastructure and toolset to advance nanotechnology.
4) Support responsible development of nanotechnology.

For each of the goals, the plan identifies specific objectives toward collectively achieving the NNI vision. Section 4 provides details about these objectives and related NNI activities.

The plan also lays out new R&D investment categories. These five Program Component Areas (PCAs), listed below, include research and development activities that contribute to one or more of the NNI goals:

1) Nanotechnology Signature Initiatives.
2) Foundational Research.
3) Nanoscale-Enabled Applications, Devices, and Systems.
4) Research Infrastructure and Instrumentation.
5) Environment, Health, and Safety.

Nanotechnology Signature Initiatives (NSIs), which are designed to accelerate innovation in areas of national priority through enhanced interagency coordination and focused investment, are called out as a priority in the 2015 OMB/OSTP R&D Priorities Memo.[5] Accordingly, they are also called out as specific NNI budget categories (PCAs) for the first time in the 2015 Budget. The NSI topics for 2013 through 2015 are as follows:

- Nanotechnology for Solar Energy Collection and Conversion: Contributing to Energy Solutions for the Future.
- Sustainable Nanomanufacturing: Creating the Industries of the Future.
- Nanoelectronics for 2020 and Beyond.
- Nanotechnology Knowledge Infrastructure (NKI): Enabling National Leadership in Sustainable Design.
- Nanotechnology for Sensors and Sensors for Nanotechnology: Improving and Protecting Health, Safety, and the Environment.

The NNI R&D investment is also guided by the 2011 NNI Environmental, Health, and Safety Research Strategy.[6] The strategy supports all four NNI goals but is most closely aligned with NNI Goal 4 and PCA 5.

The NNI funding represents the sum of the nanotechnology-related funds allocated by each of the participating agencies (the "NNI budget crosscut"). Each agency separately determines its budget for nanotechnology R&D in coordination with the Office of Management and Budget (OMB), the Office of Science and Technology Policy (OSTP), and Congress. The NNI agencies participating in the budget crosscut work closely with each other to create an integrated scientific program. This close communication, facilitated through the NSET Subcommittee and its working groups and coordinators, has led to interagency coordination and collaboration in a variety of forms, including sharing of knowledge and expertise; joint sponsorship of solicitations and workshops; and leveraging of funding, staff, and equipment/facility resources at NNI participating agencies.

Purpose of this Report

This document provides supplemental information to the President's 2015 Budget and serves as the Annual Report on the NNI called for in the 21st Century Nanotechnology Research and Development Act (P.L. 108-153, 15 USC §7501). The report also addresses the requirement for Department of Defense reporting on its nanotechnology investments, per 10 USC §2358 (see Appendix A). In particular, the report summarizes NNI programmatic activities for 2013 and 2014, as well as those currently planned for 2015. NNI budgets for 2013 through 2015 are presented by agency and PCA in Section 2 of this report. Information on the use of the Small Business Innovation Research (SBIR) and Small Business Technology Transfer Research (STTR) program funds to support nanotechnology research and commercialization activities, also called for in P.L. 108-153, is included at the end of Section 2. Section 3 outlines changes in balance of investments by PCA (including how agencies are allocating investments among the new PCAs outlined in the 2014 NNI Strategic Plan). Activities that have been undertaken and progress that has been made toward achieving the four goals set out in the NNI Strategic Plan (including activities in support of the NNI Nanotechnology Signature Initiatives) are outlined in Section 4. Highlights from external reviews of the NNI and how their recommendations are being addressed are summarized in Section 5. Appendices include supplementary information on nanotechnology investments by the Department of Defense, a list of abbreviations and acronyms, and a contact list of NSET Subcommittee participants and NNCO staff members.

2. NNI Investments

Budget Summary

The President's 2015 Budget provides over $1.5 billion for the National Nanotechnology Initiative (NNI), a continued investment in support of the President's priorities and innovation strategy. Cumulatively totaling nearly $21 billion since the inception of the NNI in 2001 (including the 2015 request), this support reflects nanotechnology's potential to significantly improve our fundamental understanding and control of matter at the nanoscale and to translate that knowledge into solutions for critical national issues. NNI research efforts are guided by two strategic documents developed by the Nanoscale Science, Engineering, and Technology (NSET) Subcommittee of the National Science and Technology Council (NSTC), the 2014 NNI Strategic Plan (www.nano.gov/2014StrategicPlan) and the 2011 NNI Environmental, Health, and Safety Research Strategy (www.nano.gov/2011EHSStrategy). These strategic documents guide how NNI agencies address the full range of nanotechnology research and development, technology transfer and product commercialization, infrastructure, and education, as well as the societal issues that accompany an emerging technology. The NNI investments in 2013 and 2014 and those proposed for 2015 continue the emphasis on accelerating the transition from basic R&D to innovations that support national priorities.

One significant change for the 2015 Budget, which is reflected in the figures provided in this document for 2013 and 2014, is a revision in the Program Component Areas (PCAs), budget categories under which the NNI investments are reported. Note that this represents an update of how NNI investments by the Federal Government are tabulated, but not a change in the overall scope of the Initiative. As outlined in the 2014 NNI Strategic Plan, the new PCAs are more broadly strategic, fully inclusive, and consistent with Federal research categories, while correlating well with the NNI goals and high-level objectives. Of particular note is the creation of a separate PCA for the Nanotechnology Signature Initiatives (NSIs), reflecting the high priority placed on NSIs in the 2015 OMB/OSTP R&D Priorities Memo.[7]

Table 2. NNI Budget, by Agency, 2013–2015 (dollars in millions)

Agency	2013 Actual	2014 Estimated*	2015 Proposed
CPSC	1.3	2.0	2.0
DHS	14.0	24.0	32.4
DOC/NIST	91.4	97.8	82.6
DOD	170.1	175.9	144.0
DOE**	314.2	303.3	343.1
DOT/FHWA	2.4	2.0	1.5
EPA	14.6	15.5	16.8
DHHS (total)	485.4	469.5	469.6
FDA	16.1	17.0	17.0
NIH	458.8	441.5	441.5
NIOSH	10.5	11.0	11.1
NASA	16.4	17.9	13.7
NSF	421.0	410.6	412.4
USDA (total)	19.5	19.1	18.8
ARS	2.0	2.0	2.0
FS	5.0	4.0	4.0
NIFA	12.5	13.1	12.8
TOTAL***	1550.2	1537.5	1536.9

* 2014 numbers are based on 2014 enacted levels, and may shift as operating plans are finalized.

** Funding levels for DOE include the combined budgets of the Office of Science, the Office of Energy Efficiency and Renewable Energy (EERE), the Office of Fossil Energy, and the Advanced Research Projects Agency for Energy (ARPA-E).

*** In Tables 2–6, totals may not add, due to rounding.

The President's 2015 Budget supports nanoscale science, engineering, and technology R&D at 11 agencies. Federal organizations with the largest investments are:

- DHHS/NIH (nanotechnology-based biomedical research at the intersection of life and physical sciences).
- NSF (fundamental research and education across all disciplines of science and engineering).
- DOE (fundamental and applied research providing a basis for new and improved energy technologies).
- DOD (science and engineering research advancing defense and dual-use capabilities).
- DOC/NIST (fundamental research and development of measurement and fabrication tools, analytical methodologies, metrology, and standards for nanotechnology).

Other agencies and agency components investing in mission-related nanotechnology research are DHS, DHHS/FDA, EPA, NASA, USDA/NIFA, DHHS/NIOSH, USDA/FS, CPSC, USDA/ARS, and DOT/FHWA.[8]

Table 2 presents NNI investments for 2013 through 2015 for Federal agencies with budgets and investments for nanotechnology R&D, including funding for the NSIs. Tables 3–5 list the investments by agency and by Program Component Area for 2013 through 2015. Table 6 shows the NNI investments within Small Business Innovation Research (SBIR) and Small Business Technology Transfer (STTR) programs for 2008 through 2012.

Key Points About the 2013–2015 NNI Investments

- NNI investments for 2013 through 2015 are reported here for the first time under the updated PCA categories called out in the 2014 NNI Strategic Plan. Investments under these revised categories should not be compared directly to investments in the previous PCAs set out in prior NNI strategic plans and reported in previous NNI budget supplement reports, with the exception of the environmental, health, and safety (EHS) PCA, which remains unchanged between the old and new PCA categories.
- 2013 actual NNI investments ($1.55 billion) reported in this document are approximately 6% less than the estimates previously published in the NNI Supplement to the President 2014 Budget. This difference is consistent with the budget reduction associated with the sequester for 2013, which was enacted after the publication of that supplement.

Table 3. Actual 2013 Agency Investments by Program Component Area (dollars in millions)

	1. Nanotechnology Signature Initiatives (NSIs)*	Solar Energy	Nanomanufacturing	Nanoelectronics	NKI	Sensors	2. Foundational Research	3. Nanotechnology-Enabled Applications, Devices, and Systems	4. Research Infrastructure and Instrumentation	5. Environment, Health, and Safety	NNI Total
CPSC	**0.0**	0.0	0.0	0.0	0.0	0.0	**0.0**	**0.0**	**0.0**	**1.3**	**1.3**
DHS	**0.0**	0.0	0.0	0.0	0.0	0.0	**0.0**	**5.9**	**0.0**	**8.1**	**14.0**
DOC/NIST	**29.2**	3.1	3.0	18.2	1.1	3.9	**13.3**	**7.3**	**35.4**	**6.2**	**91.4**
DOD	**51.0**	4.2	3.7	26.3	1.7	15.0	**76.2**	**38.3**	**1.5**	**3.1**	**170.1**
DOE	**44.0**	37.0	0.0	0.0	4.5	2.5	**131.0**	**27.4**	**111.8**	**0.0**	**314.2**
DOT/FHWA	**0.0**	0.0	0.0	0.0	0.0	0.0	**0.0**	**2.4**	**0.0**	**0.0**	**2.4**
EPA	**0.0**	0.0	0.0	0.0	0.0	0.0	**0.0**	**0.0**	**0.0**	**14.6**	**14.6**
DHHS (total)	**51.0**	0.0	0.0	0.0	0.0	51.0	**153.3**	**214.7**	**16.6**	**49.8**	**485.4**
FDA	0.0	0.0	0.0	0.0	0.0	0.0	0.0	0.0	0.0	16.1	16.1
NIH	51.0	0.0	0.0	0.0	0.0	51.0	153.3	214.7	16.6	23.2	458.8
NIOSH	0.0	0.0	0.0	0.0	0.0	0.0	0.0	0.0	0.0	10.5	10.5
NASA	**5.0**	0.0	3.2	0.2	0.3	1.3	**6.4**	**4.6**	**0.4**	**0.0**	**16.4**
NSF	**93.6**	28.4	22.7	42.6	0.0	0.0	**196.8**	**55.9**	**43.7**	**31.0**	**421.0**
USDA (total)	**6.2**	1.0	2.2	0.0	0.0	3.0	**4.3**	**4.9**	**3.0**	**1.1**	**19.5**
ARS	0.0	0.0	0.0	0.0	0.0	0.0	0.0	2.0	0.0	0.0	2.0
FS	2.0	1.0	1.0	0.0	0.0	0.0	2.0	0.0	1.0	0.0	5.0
NIFA	4.2	0.0	1.2	0.0	0.0	3.0	2.3	2.9	2.0	1.1	12.5
TOTAL	**279.9**	73.6	34.7	87.3	7.5	76.8	**581.3**	**361.4**	**212.5**	**115.1**	**1550.2**

* Abbreviated titles for the Nanotechnology Signature Initiatives are shown in Tables 3–5. See the text (Chapter 1, p. 5) for full NSI titles.

- The Foundational Research PCA constitutes over a third of the total NNI investment portfolio for 2013 through 2015. The NNI agencies view this as a critical investment category, maintaining a pipeline of new nanotechnology-based innovations, even as prior foundational research matures into applications.
- Investments in Nanotechnology Signature Initiatives (PCA 1) have grown since their inception, from approximately $246 million in 2011[9] to a request of over $291 million in the 2015 Budget. NSIs now represent nearly 20% of the total NNI investments for 2013 through 2015. EHS research investments have been sustained at approximately $110 million per year, reflecting their continued high priority for the NNI. EHS investments have risen from under 3% of the total NNI investment in 2005 to over 7% in the 2015 Budget. Cumulative EHS investments from 2005 through 2015 have now reached over $900 million.

Table 4. Estimated 2014 Agency Investments by Program Component Area (dollars in millions)

	1. Nanotechnology Signature Initiatives (NSIs)	Solar Energy	Nanomanufacturing	Nanoelectronics	NKI	Sensors	**2. Foundational Research**	**3. Nanotechnology-Enabled Applications, Devices, and Systems**	**4. Research Infrastructure and Instrumentation**	**5. Environment, Health, and Safety**	**NNI Total**
CPSC	**0.0**	0.0	0.0	0.0	0.0	0.0	**0.0**	**0.0**	**0.0**	**2.0**	**2.0**
DHS	**0.0**	0.0	0.0	0.0	0.0	0.0	**0.0**	**16.1**	**0.0**	**7.9**	**24.0**
DOC/NIST	**29.2**	3.3	5.6	17.3	1.0	2.1	**15.3**	**8.7**	**39.5**	**5.1**	**97.8**
DOD	**50.8**	2.6	1.0	24.8	5.1	17.2	**84.8**	**37.0**	**0.1**	**3.2**	**175.9**
DOE	**42.3**	35.3	0.0	0.0	5.0	2.0	**129.0**	**16.1**	**115.8**	**0.0**	**303.3**
DOT/FHWA	**0.0**	0.0	0.0	0.0	0.0	0.0	**0.0**	**2.0**	**0.0**	**0.0**	**2.0**
EPA	**0.0**	0.0	0.0	0.0	0.0	0.0	**0.0**	**0.0**	**0.0**	**15.5**	**15.5**
DHHS (total)	**49.1**	0.0	0.0	0.0	0.0	49.1	**119.0**	**228.1**	**19.9**	**53.3**	**469.5**
FDA	0.0	0.0	0.0	0.0	0.0	0.0	0.0	0.0	0.0	17.0	**17.0**
NIH	49.1	0.0	0.0	0.0	0.0	49.1	119.0	228.1	19.9	25.3	**441.5**
NIOSH	0.0	0.0	0.0	0.0	0.0	0.0	0.0	0.0	0.0	11.0	**11.0**
NASA	**7.9**	0.3	5.0	0.1	0.1	2.4	**6.7**	**3.3**	**0.0**	**0.0**	**17.9**
NSF	**112.9**	25.2	24.7	34.5	20.9	7.6	**180.6**	**46.9**	**44.9**	**25.3**	**410.6**
USDA (total)	**5.9**	0.7	2.1	0.0	0.0	3.1	**4.3**	**4.9**	**3.0**	**1.0**	**19.1**
ARS	0.0	0.0	0.0	0.0	0.0	0.0	0.0	2.0	0.0	0.0	**2.0**
FS	1.0	0.0	1.0	0.0	0.0	0.0	2.0	0.0	1.0	0.0	**4.0**
NIFA	4.9	0.7	1.1	0.0	0.0	3.1	2.3	2.9	2.0	1.0	**13.1**
TOTAL	**298.1**	67.4	38.4	76.7	32.1	83.5	**539.8**	**363.2**	**223.2**	**113.3**	**1537.5**

- Nanotechnology investments have remained stable at most NNI participating agencies since 2010, even with many competing priorities. This stability demonstrates the continued competitiveness of nanotechnology R&D in supporting agency missions.
- The Department of Defense has experienced a substantial reduction in its investments since 2012. DOD funding actuals for 2013, which are approximately 50% lower than 2012 actuals, reflect the completion of projects, especially at DARPA, where investments (particularly in nanotechnologyenabled devices and systems) have matured. DOD investments in foundational research have been sustained. Because nanotechnology investments by DOD are wholly driven by mission needs and opportunities, changes in investment are to be expected.
- The Department of Health and Human Services accounts for the largest portion of the total NNI investment, reflecting progress in the development of potentially revolutionary nanomedicine applications at NIH and a strong and sustained EHS research program at FDA, NIOSH, and NIH's National Institute of Environmental Health Sciences.

Table 5. Proposed 2015 Agency Investments by Program Component Area (dollars in millions)

	1. Nanotechnology Signature Initiatives (NSIs)	Solar Energy	Nanomanufacturing	Nanoelectronics	NKI	Sensors	2. Foundational Research	3. Nanotechnology-Enabled Applications, Devices, and Systems	4. Research Infrastructure and Instrumentation	5. Environment, Health, and Safety	NNI Total
CPSC	**0.0**	0.0	0.0	0.0	0.0	0.0	**0.0**	**0.0**	**0.0**	**2.0**	**2.0**
DHS	**0.0**	0.0	0.0	0.0	0.0	0.0	**0.0**	**24.4**	**0.0**	**8.0**	**32.4**
DOC/NIST	**24.4**	3.1	6.1	11.5	1.3	2.4	**11.3**	**4.6**	**35.6**	**6.7**	**82.6**
DOD	**47.3**	2.3	0.8	22.0	0.9	21.3	**72.2**	**22.4**	**0.0**	**2.1**	**144.0**
DOE	**42.8**	35.3	0.0	0.0	5.0	2.5	**136.7**	**29.2**	**134.4**	**0.0**	**343.1**
DOT/FHWA	**0.0**	0.0	0.0	0.0	0.0	0.0	**0.0**	**1.5**	**0.0**	**0.0**	**1.5**
EPA	**0.0**	0.0	0.0	0.0	0.0	0.0	**0.0**	**0.0**	**0.0**	**16.8**	**16.8**
DHHS (total)	**49.1**	0.0	0.0	0.0	0.0	49.1	**119.0**	**228.1**	**19.9**	**53.4**	**469.6**
FDA	0.0	0.0	0.0	0.0	0.0	0.0	0.0	0.0	0.0	17.0	**17.0**
NIH	49.1	0.0	0.0	0.0	0.0	49.1	119.0	228.1	19.9	25.3	**441.5**
NIOSH	0.0	0.0	0.0	0.0	0.0	0.0	0.0	0.0	0.0	11.1	**11.1**
NASA	**6.4**	0.1	3.9	0.0	0.0	2.4	**5.6**	**1.3**	**0.4**	**0.0**	**13.7**
NSF	**115.6**	27.7	23.4	38.0	19.0	7.5	**181.4**	**46.6**	**46.5**	**22.3**	**412.4**
USDA (total)	**5.7**	0.7	2.0	0.0	0.0	3.0	**4.2**	**4.9**	**3.0**	**1.0**	**18.8**
ARS	0.0	0.0	0.0	0.0	0.0	0.0	0.0	2.0	0.0	0.0	**2.0**
FS	1.0	0.0	1.0	0.0	0.0	0.0	2.0	0.0	1.0	0.0	**4.0**
NIFA	4.7	0.7	1.0	0.0	0.0	3.0	2.2	2.9	2.0	1.0	**12.8**
TOTAL	**291.3**	69.2	36.2	71.5	26.2	88.2	**530.4**	**363.0**	**239.8**	**112.4**	**1536.9**

- The Department of Homeland Security is proposing to more than double its nanotechnology investments, from $14 million in 2013 to over $32 million in the 2015 Budget.
- As in several recent years, NNI agencies are also actively participating in and contributing to other complementary and synergistic Administration R&D priorities, including the Networking and Information Technology Research and Development (NITRD) Program, the Global Change Research Program (GCRP), the Materials Genome Initiative (MGI), Advanced Manufacturing, and the Brain Research through Advancing Innovative Neurotechnologies (BRAIN) Initiative. Nanoscale science and technology is related to, enables, or is enabled by each of these other priority topics in various ways, and NNI investments may overlap with investments reported for each of them. NNI agency representatives and NNCO staff continue to explore opportunities to take full advantage of these synergies.
- Investments in SBIR and STTR funding by the participating agencies, reported outside of the formal NNI funding crosscut and the budget tables shown above, play a critical role in transitioning nanotechnology innovations into products for commercial and public benefit (NNI Goal 2), as discussed below.

Utilization of SBIR and STTR Programs to Advance Nanotechnology

As called for by the 21st Century Nanotechnology Research and Development Act, this report includes information on use of the Small Business Innovation Research and Small Business Technology Transfer programs in support of nanotechnology development. Five NNI agencies—DOD, NSF, NIH, DOE, and NASA—have both SBIR and STTR programs. In addition, NIOSH, EPA, USDA, and NIST have SBIR programs. Table 6 shows agency funding for SBIR and STTR awards for nanotechnology R&D from 2008 through 2012 (the latest year for which data are available).

Some NNI agencies (e.g., NSF and NIH) have included nanotechnology-specific topics in their SBIR and STTR solicitations. The NSF SBIR program has an ongoing nanotechnology topic included within its advanced materials, electronics, and biotechnology solicitations. Some agencies have had additional topical or applications-oriented solicitations for which many awardees have proposed nanotechnology-based innovations. SBIR/STTR data for 2004 through 2012 indicates that the NNI agencies have funded over $800 million of nanotechnology-related SBIR and STTR awards. Data from 2008 through 2012 are shown in Table 6, and a complete listing by year (including data from 2004 through 2007) can be found at a download link in the Overview section of nanodashboard.nano.gov.

Table 6. 2008–2012 Agency SBIR and STTR Awards (dollars in millions)

	2008			2009			2010			2011			2012		
	SBIR	STTR	Total	SBIR	STTR	Total	SBIR	STTR	Total	SBIR	STTR	Total	SBIR	STTR	Total
DOD	19.8	2.3	22.1	19.3	3.9	23.2	32.2	10.4	42.6	16.2	5.6	21.8	25.4	7.6	33.0
NSF	10.5	7.5	18.0	13.2	7.0	20.2	25.2	3.9	29.2	13.2	1.3	14.5	18.0	2.5	20.5
DHHS/NIH	29.3	1.8	31.1	23.0	3.7	26.7	21.7	1.6	23.2	22.6	1.5	24.1	21.7	3.9	25.6
DHHS/ NIOSH	0.4	0.0	0.4	0.7	0.0	0.7	0.0	0.0	0.0	0.2	0.0	0.2	0.0	0.0	0.0
DOE[10]	13.8	2.7	16.5	31.9	5.7	37.6	19.4	4.4	23.8	4.7	1.1	5.8	7.8	1.7	9.5
NASA	6.2	0.8	7.0	15.0	1.7	16.7	11.6	2.1	13.7	12.1	3.7	15.8	17.0	3.9	20.8
EPA	0.7	0.0	0.7	0.7	0.0	0.7	0.4	0.0	0.4	0.9	0.0	0.9	0.7	0.0	0.7
USDA[11]	0.9	0.0	0.9	0.5	0.0	0.5	0.0	0.0	0.0	1.0	0.0	1.0	0.5	0.0	0.5
DOC/NIST	0.4	0.0	0.4	0.4	0.0	0.4	0.2	0.0	0.2	0.3	0.0	0.3	0.4	0.0	0.4
TOTAL	**81.9**	**15.1**	**97.0**	**104.6**	**22.0**	**126.6**	**110.6**	**22.4**	**133.0**	**71.2**	**13.2**	**84.4**	**91.4**	**19.6**	**111.0**

3. Changes in Balance of Investments by Program Component Area

The 21st Century Nanotechnology R&D Act calls for this report to address changes in the balance of investments by NNI member agencies among the Program Component Areas (PCAs). Because the NNI PCAs have changed significantly with the publication of the updated 2014 NNI Strategic Plan,[12] some text is included in this section for all of the agencies that are part of the NNI funding crosscut to explain how they are reporting their investments under the revised PCAs.

CPSC: The investments in nanotechnology for CPSC for 2013 and 2014 have focused on addressing the environmental, health, and safety issues identified in PCA 5. Data produced in these research efforts may also support the Nanotechnology Knowledge Infrastructure (NKI) Signature Initiative described in PCA 1. It is expected that research investments for 2015 will continue to support PCA 5.

DHS: The Department of Homeland Security is reporting investments for 2013 through 2015 in PCAs 3 (Nanotechnology-Enabled Applications, Devices, and Systems) and 5 (Environment, Health, and Safety). DHS is requesting a significant increase in PCA 3 investments for homeland security applications, from $5.9 million in 2013 to $24.4 million in the 2015 request. See DHS text in Section 4 (Individual Agency Contributions to Objective 1.1) for additional details.

DOD:[13] The DOD does not establish funding targets for nanotechnology, which has proven to be competitive based on its contributions to meeting needs and providing opportunities for future capabilities. New projects are awarded on a competitive basis, and therefore, the balance of investment can change from historical levels and predictions.

In 2015 DOD anticipates about $144 million in nanotechnology research and development, reduced from 2014 ($176 million), and about 15% less than 2013 actual expenditures. Actuals tend to be larger than estimates due to success in competitive solicitations. The revised PCA structure provides a more meaningful and illustrative means of representing the broad nature of the DOD investment. The traditional 50:50 breakdown between fundamental and applied research is approximately retained, but the resources are now more definitively characterized. About 45% of the DOD nanotechnology-related investment is in PCA 2 (Foundational Research)—rising to 50% in the 2015 request. About 23% is in PCA 3 (Nanotechnology-Enabled Applications, Devices, and Systems)—dropping to 16% in the 2015 request. Both of these changes are due in part to completion of projects, particularly at DARPA, in applied or applications-directed research classified under PCA 3. About 30% of the DOD investment is currently in the areas of the signature initiatives and clearly shows the topical emphasis on nanoelectronics (15%) and sensors (also 15% in the 2015 request). The projects associated with nanomanufacturing in 2013 and 2014 are completing, but the emphasis on this area of research remains strong and will continue to be an area of competitive opportunity. The DOD Small Business Innovation Research (SBIR) and Small Business Technology Transfer (STTR) investments in nanotechnology remain very strong with an overall investment of $33 million in 2012, making these projects about 20% of the overall DOD nanotechnology research investment.

The revision in PCAs is largely responsible for changes in distribution from prior years. In 2013–2015 funding, the primary changes for DOD are associated with fluctuations in the signature initiatives and nanotechnology-enabled PCAs as projects complete and new opportunities for competition present themselves.

DOE: In 2015, increases in DOE funding in PCA 4 are due primarily to the planned merger of three Electron-Beam Microcharacterization Centers (EBMCs) with the Nanoscale Science Research Centers (NSRCs), which will enhance the availability and use of state-of-the-art microscopy tools for nanoscience. In addition, DOE has requested an approximately 5% increase in the NSRC budgets for 2015 to allow for improved operational levels.

DOT/FHWA: Under the current PCA construct, all of FHWA's current and planned investment is in Nanotechnology-Enabled Applications, Devices, and Systems (PCA 3). FHWA's investment seeks to leverage developments in nanotechnology by focusing them on transportation applications such as corrosion protection, lightweight structural and multifunctional materials, and *in situ* and onboard sensing and communications devices.

EPA: The EPA 2015 Budget request continues EPA's strong support for research in environmental, health, and safety applications of nanotechnology. EPA will continue work in PCA 5 on evaluating the chemical and physical properties of engineered nanomaterials that influence their potential for release into the environment; fate, transport, and transformation; and potential for exposure and adverse effects to humans or sensitive ecological species. The potential for nanomaterial releases will be evaluated across the product life cycle including manufacturing, product use, and end-of-life disposal. Research includes the mechanisms of interactions between engineered nanomaterials and biological organisms, which has been identified by the National Academies as a high priority.

FDA: FDA continues to focus on PCA 5 (Environment, Health, and Safety) with no change in investment balance. FDA investments in PCA 5 will continue to enable the agency to address products containing nanomaterials or involving the use of nanotechnology through the development of methods, validation of models, and generation of data that are necessary for safety and efficacy assessments of nanotechnologyenabled products.

NASA: NASA support for nanotechnology R&D has been focused primarily on PCA 1 (Nanotechnology Signature Initiatives), PCA 2 (Foundational Research), and PCA 3 (Nanotechnology Enabled Applications, Devices, and Systems). Nanotechnology R&D has been supported primarily by the Aeronautics Research Mission Directorate (PCA 3) and the Space Technology Mission Directorate (PCAs 1, 2, and 3). NASA investments in 2015 in PCA 1 are expected to grow by almost 30% relative to 2013 due to increased funding in nanostructured materials and nanotechnology-based chemical-biological sensors. Research investments in PCA 2 are expected to remain largely unchanged in 2015, while requested investments in PCA 3 are reduced in 2015 from 2013 levels.

NIH: NIH currently invests approximately $450 million per year in nanotechnology-related biomedical research, which is expected to remain relatively constant for 2015 based on current grant funding policies and programmatic initiatives. For example, in 2013, there were five active nanotechnology-specific funding opportunity announcements (FOAs) and four major investment programs supported by the various NIH institutes that contributed to the total dollar investment. The major areas of research activity from both unsolicited and solicited resources focused on PCAs 1, 2, and 3 with more than 1,200 grants funded. Since then, the two SBIR/STTR FOAs have expired and were not renewed as stand-alone FOAs, given the increase in nanotechnology-related activities supported through other existing SBIR/STTR initiatives. The renewal of the National Cancer Institute's (NCI) Provocative Questions FOA also led to increased support for primarily PCAs 2 and 3. Overall, research projects funded in 2014 through both the investigator-initiated and programmatic initiatives are focused primarily on PCA 3, Nanotechnology-Enabled Applications, Devices, and Systems (51.7%) for biomedical research, and to a lesser extent on PCA 4, Research Infrastructure and Instrumentation (4.5%), and PCA 5 (5.7%). In addition, the four major NIH nanotechnology programs, specifically, the NIH Common Fund Nanomedicine Initiative, NCI's Alliance for Nanotechnology in Cancer, the National Heart, Lung and Blood Institute's (NHLBI) Program of Excellence in Nanotechnology, and the National Institute of Environmental Health Sciences (NIEHS) Centers for Nanotechnology Health Implications Research initiated prior to 2014 continued to receive support to develop new treatments, diagnostics, and interventions for disease management and patient care related to nanotechnology. As a result, the amount invested per PCA has not changed significantly from 2013 to 2014 due to the continuation of research and the awarding of new grants.

The NIH investment in nanotechnology in 2015 will continue to support the ongoing research projects within these programmatic initiatives as well as the Nanotechnology Signature Initiatives. Emphasis at this stage of development will include new clinically relevant diagnostics and devices, novel uses of clinical imaging and drug delivery technologies, and advanced genetic sequencing devices/assays for biological profiling (impacting PCAs 1, 2, and 3). NIH will also maintain resources toward achieving the goals of the 2011 EHS Research Strategy and continue to lead the Sensors NSI and participate in the Nanotechnology Knowledge Infrastructure NSI (see Section 4 for defined research activity).

New areas of opportunities for nanotechnology are also expected to arise in 2015 as the institutes across the NIH implement new programs/initiatives/plans to achieve the scientific vision of President Obama's Brain Research through Advancing Innovative Neurotechnologies (BRAIN) Initiative. The Initiative promises to open new doors to understanding how the brain functions and is linked to mechanisms of brain disease by producing real-time pictures of complex neural circuits and visualizing the rapid-fire interactions of cells that occur at the speed of thought. The goal is to build on the growing scientific foundations in neuroscience, genetics, physics, engineering, informatics, nanoscience, chemistry, mathematics, and other advances of the past few decades to catalyze an interdisciplinary effort of unprecedented scope. Nonetheless, the total dollar investment in this area for 2015 is not expected to significantly change relative to previous years due to ongoing funding and limited programmatic changes.

NIH's investments in nanotechnology from 2013 to 2015 will not result in significant redistribution of funds across the PCAs due to continuation of existing initiatives, programs, and grant research. However, the actual NIH investments reported for 2015 in the future may change as NIH continues to refine and validate its reporting system to the updated PCA classification scheme.

NIOSH: NIOSH support for nanotechnology R&D is reported under PCA 5 (Environment, Health, and Safety), although a number of its activities also are related to and support the Nanotechnology Signature Initiatives (PCA 1). In 2015, NIOSH will maintain an active field investigation effort in support of the Sustainable Nanomanufacturing NSI that focuses on partnering with private sector nanomanufacturing facilities and providing comprehensive risk management guidance that promotes responsible development and supports speed to market. A critical focus area in 2015 will be the development and implementation of engineering control strategies and solutions that mitigate nanomaterials process releases and worker exposures. This program will continue to maintain focus in two areas: providing guidance to nanomanufacturing that is using established materials such as titania and carbon nanotubes and developing prospective guidance for emerging materials such as cellulose nanocrystals and graphene by conducting hazard assessment employing *in vitro* and *in vivo* models of exposure-response relationships. For 2015, NIOSH will continue its agency shift of emphasis within its program to fill knowledge gaps in five strategic areas:

- Increase the understanding of nanomaterials' new human health hazards and health risks to workers.

- Expand understanding of the initial hazard findings on engineered nanomaterials, including development of biomarkers of exposure and early response to support worker surveillance efforts.
- Continue the development and dissemination of high-impact guidance materials that inform workers, employers, regulatory agencies, and decision makers about hazards, risks, and risk management solutions.
- Support epidemiologic studies to understand the scope of worker exposure and potential health impacts.
- Promote and assess national adherence to risk management guidance.

NIOSH will continue to build partnerships with private sector nanomaterial companies and will coordinate efforts with other NNI agencies, including OSHA, NIST, CPSC, EPA, and DOD. An updated 2013–2016 NIOSH Nanotechnology Research Strategic Plan[14] describes the new areas of emphasis.

NIST: NIST's intramural research laboratories, user facilities, and services work at the cutting edge of science to ensure that U.S. companies, as well as the broader science and engineering communities, have the nanoscale measurements, data, and technologies to further nanotechnology innovation and industrial competitiveness. The 2015 Budget request continues NIST's strong support for research in nanotechnology. The request continues a trajectory of increased investment in the key NIST priority area of nanomanufacturing (PCA 1, Nanotechnology Signature Initiative for Sustainable Nanomanufacturing). NIST's commitment to nanotechnology-related environment, health, and safety research (PCA 5) is sustained, and the considerable investment begun in 2012 for major research facilities and instrumentation acquisition (PCA 4) remains NIST's largest nanotechnology investment area. In 2015 some instrumentation investments are reported in the Nanotechnology Signature Initiative (PCA 1) that the requested funds are intended to directly support. Overall funding amounts have decreased in many areas because of a revision in the approach to categorizing research planned for 2015, combined with changes in investments associated with significant program turnover and the transition of activities within NIST toward advanced manufacturing investments including precision measurements and advanced materials design and discovery.

NSF: NSF supports nanoscale science and engineering throughout all the research and education directorates as a means to advance discovery and innovation and integrate various fields of research. The NNI enables increased interdisciplinarity through research at atomic and molecular levels for about 5,000 active awards with full or partial contents related to nanoscale science and engineering (NSE).

The NSF 2015 request will have a decrease of about 2% as compared to the 2013 actuals. The support for Nanotechnology Signature Initiatives (PCA 1) will grow from $93.6 million in 2013 to $115.6 million (a 23.5% increase) in the 2015 request, with significant increases for the Nanotechnology Knowledge Infrastructure NSI and the Sensors NSI, to $19 million and $7.5 million, respectively. The investment in Research Infrastructure and Instrumentation (PCA 4) will also grow, whereas Foundational Research (PCA 2); Nanotechnology-Enabled Applications, Devices, and Systems (PCA 3); and EHS (PCA 5) investments will decrease. Those decreases are caused in part by the results of the peer review process of unsolicited proposals and co-funding available from other NNI agencies. Part of PCA 3 will be dedicated to research on breakthrough materials and advanced manufacturing as part of the NSF-wide

Cyber-Enabled Materials, Manufacturing, and Smart Systems (CEMMSS) investment framework.

Three Nanosystems Engineering Research Centers (NERCs), with a total estimated budget of approximately $55 million for five years, were established in September 2012 and started full operation in 2013 and will be expanded by 2015. Additional NERCs are under competition.

A new competition in 2014 is expected to yield additional NERCs. In 2015 the agency continues its contributions to translational innovation programs, including GOALI (Grant Opportunities for Academic Liaison with Industry), I/UCRC (Industry/University Cooperative Research Centers), the I-Corps (Innovation Corps program), AIR (Accelerating Innovation Research), PFI (Partnerships for Innovation), and PFI's Building Innovation Capacity (BIC) component. The NSF Small Business Innovation Research program has an ongoing nanotechnology topic with subtopics for nanomaterials, nanomanufacturing, nanoelectronics and active nanostructures, nanotechnology for biological and medical applications, and instrumentation for nanotechnology.

NSF has mainstreamed nanotechnology-related research in core programs in several directorates. There is a continuing strong interest from the community in nanotechnology research, education, and infrastructure through those core programs.

Nanoscale Science and Engineering is substantially represented in NSF-wide activities in 2014 and 2015 such as Science, Engineering, and Education for Sustainability/Sustainable Chemistry, Engineering and Materials (SusChEM); Research at the Interface of Biological, Mathematical and Physical Sciences, and Engineering (BioMaPS); and Smart Systems (CEMMSS), which includes Designing Materials to Revolutionize and Engineer our Future (DMREF) and Understanding the Brain. NSF's Directorates for Mathematical and Physical Sciences (MPS), Engineering (ENG), and Computer and Information Science and Engineering (CISE) are considering additional nanotechnology emphasis in the areas of optics and photonics in 2015.

USDA/ARS: The President's 2015 Budget request for ARS will continue to support research in nanotechnology. Classification of 2013 and 2014 investments highlights a focus on PCA 3, Nanotechnology-Enabled Applications, Devices, and Systems. An approximate comparison to 2013 levels indicates that the requests for 2015 will remain constant.

USDA/FS: Forest Service nanotechnology research supports USDA's priorities of sustaining forest health, stimulating rural development, and creating value for our Nation's landowners.

In the 2015 Budget request, the Forest Service will support research on manufacturing, characterization, and application of cellulose nanomaterials originating from operations to restore forest vitality. The majority of the Forest Service's nanotechnology R&D supported by the 2015 Budget are in the Sustainable Nanomanufacturing Signature Initiative (PCA 1) and in Foundational Research (PCA 2). The Forest Service plans to invest the balance of its nanotechnology-related request in the 2015 Budget in Research Infrastructure and Instrumentation (PCA 4). The 2015 Budget request reflects a 20% reduction in the Forest Service's total R&D investments from 2014. However, nanotechnology remains a priority within the FS R&D portfolio. Forest Service has established a public–private partnership to continue its research in nanotechnology, even under this decreased overall R&D Budget request.[15]

USDA/NIFA: The National Institute of Food and Agriculture will continue to support research in nanotechnology for agriculture and food systems. Currently, more than 300 active research and education projects in advancing nanoscale science, engineering, and technology are supported by NIFA to address grand societal challenges facing agriculture, food, health and nutrition, the environment, and rural communities. Analysis of 2013 investments reflects that NIFA has maintained a balanced approach in all five PCAs. About 20% of the investment is in PCA 2 (Foundational Research) and about 25% in PCA 3 (Nanotechnology-Enabled Applications, Devices, and Systems).

NIFA also reports a 16% investment in PCA 4 (Research Infrastructure and Instrumentation), as the new classification now includes efforts in curriculum development and human infrastructure development. More than 30% of the investment is currently in the areas of the Nanotechnology Signature Initiatives (PCA 1), with the topical emphasis on the Sensors NSI (24%) and the Sustainable Nanomanufacturing NSI (9%). NIFA is also involved in the Nanotechnology for Solar Energy Collection and Conversion NSI. NIFA maintains about 9% of its investment in Environment, Health, and Safety (PCA 5). The request for 2015 will essentially remain the same as compared to the 2013 investment level.

4. Progress Towards Achieving NNI Goals, Objectives, and Priorities

Activities Relating to the Four NNI Goals and Fifteen Objectives

As called for in the 21st Century Nanotechnology Research and Development Act, this NNI Supplement to the President's Budget provides an analysis of the progress made toward achieving the goals established for the National Nanotechnology Initiative. The NNI's activities for 2013 and 2014 and plans for 2015 are reported here in terms of how they promote progress toward the four NNI goals and fifteen objectives set out in the 2014 NNI Strategic Plan (Figure 1). It is important to note that many agency programs and activities, although listed below under their primary NNI goals and objectives, address components of multiple goals simultaneously. Therefore an integrated perspective across the four goals is needed to understand progress towards achieving the NNI vision.

Goal- and objective-related activities are reported below in terms of two categories of activities: (1) individual agency activities and (2) coordinated activities of NNI agencies with other agencies and with groups external to the NNI, including international activities. *The activities described below in terms of these two categories are only selected highlights of current and planned work of the NNI member agencies and are not an all-inclusive description of ongoing NNI activities.*

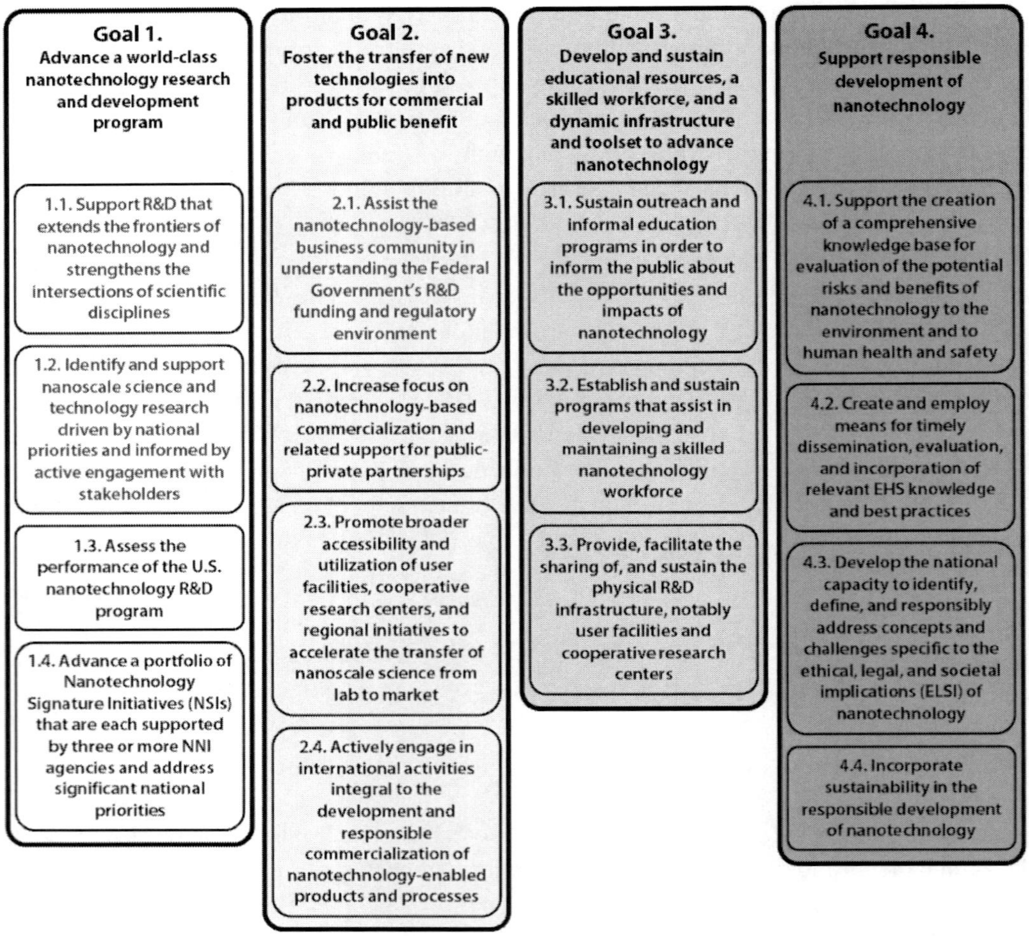

Figure 1. A summary of NNI goals and objectives from the 2014 NNI Strategic Plan.[16]

Goal 1: Advance a World-Class Nanotechnology Research and Development Program

U.S. leadership in nanotechnology R&D depends on the continued investment of participating agencies in the stimulation of discovery and innovation. These activities expand the boundaries of knowledge through a comprehensive R&D program that builds on the foundation established over the past thirteen years. The Nanotechnology Signature Initiatives (NSIs), while developed in the context of all four NNI goals, are explicitly called out in Objective 1.4 under Goal 1, as detailed in the 2014 NNI Strategic Plan. The NSIs provide a spotlight on critical areas and define the shared vision of the participating agencies to accelerate the advancement of nanoscale science and technology from research through commercialization. This enhanced coordination enables the leveraging of individual agency programs, resources, and capabilities in addition to the development of collaborative programs, and both categories are reported below.

Objective 1.1 – Support R&D That Extends the Frontiers of Nanotechnology and Strengthens the Intersections of Scientific Disciplines

Individual Agency Contributions to Objective 1.1

DHS: The DHS Science and Technology Directorate plans to continue the development of micro- and nanoscale structures for electron emitters in high-performance mass spectrometers for explosives trace detection systems and x-ray sources in next-generation baggage screening systems. The use of nanofibrous organic and polymeric materials for vapor sensing of explosives is also under investigation for potential applications in handheld explosives trace detection systems. The current focus is on development and commercialization of these component technologies into original equipment manufacturer (OEM) systems for DHS deployment.

DOD: The DOD nanoscience and foundational research efforts are associated with about half of the department's nanotechnology resources. Nanoscience and processing of materials at the nanoscale are important aspects for increasing the specific performance of materials and items upon which future defense capability can be built. Please refer to Appendix A for a detailed account of plans and progress associated with this objective.

DOE: The DOE Office of Science (DOE-SC) supports a broad and diverse investment in nanoscience research and development, primarily through grants to individual researchers at universities, funding for research groups at DOE laboratories, and support of interdisciplinary efforts such as the Energy Frontier Research Centers (EFRCs).

IC/DNI: The National Reconnaissance Office (NRO) continues to work on enhancing the properties of carbon nanotubes (CNTs), specifically in three critical areas:

- Better chirality control and separation of various types of CNTs for improved performance.
- Better alignment of CNTs during growth; initial alignment has been proven, and post-processing alignment experiments are being conducted.
- Increased density of CNTs in sheets and yarns; dramatic improvements of five- to ten-fold have been achieved through use of super fuels and will enter commercialization ramp-up in 2014, thus resulting in reduced cost and increased quantities being manufactured.

Ongoing research into decreasing direct current (DC) resistivity for 48-volt DC data center operations has now achieved one-twelfth the resistivity of copper wire in the lab and continues to make great strides. The goal is to save 10–15% energy consumption in every U.S. data center. Nonvolatile carbon nanotube-based memory is scheduled for market trials in 2014. The goal is to eliminate traditional synchronous random access memory (SDRAM), flash memory buffers, and hard disk drives from data centers for savings in size, weight, performance, and cost. This will allow a 100-fold speed-up in data processing.

Academia continues to tackle the challenge of finding a suitable nano-resin for composites that is optimized for CNTs, resulting in superior strength and enhanced acoustic dampening.

New academic research into development of boron nitride nanotubes will commence in early 2014. This is the perfect insulator to match CNT materials. NRO continues to make progress in developing superior-strength CNT bulk materials with improved electrical and thermal conductivity.

NASA: NASA has continued to support both foundational and applied research in nanotechnology at NASA centers, at universities, and in industry. For example, the Space Technology Research Fellowship Program funded 12 graduate fellowships in 2013 and a total of 30 fellowships since its inception in 2011 to perform foundational research in nanotechnology-related areas in collaboration with NASA scientists and engineers. In-house research through the Space Technology Mission Directorate's Center Innovation Funds and the Aeronautics Mission Research Directorate's Seedling Fund has been supported in topics such as the development of nanoscale coatings to improve the aerodynamic efficiency of aircraft, carbon nanotube- and boron nitride nanotube-based materials and devices, nanoscale vacuum electronics, and high-bandwidth antennas utilizing a polyimide aerogel substrate.

NIH: NIH advances world-class nanotechnology research through its extensive portfolio of extramural awards, awarded in response to both investigator-initiated applications and NIH-directed requests for applications. Investigator-initiated research largely focuses on nanoscale phenomena and development of nanomaterials and nanotechnology-enabled devices for biomedical applications. NIH investment directed specifically towards nanotechnology focuses heavily on the translation of discoveries from basic research into clinically applicable technologies, including materials for biomedical imaging and drug delivery and devices for disease detection and monitoring. This emphasis on translational research brings together the basic nanotechnology, engineering, and biomedical and clinical research communities.

The NIH institutes and centers play a significant role in nanotechnology research and development. Due to the diversity and comprehensiveness of research supported by NIH, the following highlights are limited to certain investment programs or new opportunities that demonstrate how NIH activities are advancing the frontiers of nanotechnology across disciplines. For instance, the NCI Alliance for Nanotechnology in Cancer, the National Heart, Lung and Blood Institute (NHLBI) Programs in Excellence in Nanotechnology, and the NIH Nanomedicine Initiative will continue to push the boundaries of biomedical nanotechnology R&D. The NCI Alliance has undergone programmatic reviews by the NCI Scientific Program Leaders and the Institute's Board of Scientific Advisors and was approved to continue a third phase starting at the end of 2015. Phase III will last five years and be populated by a request for applications for funds set aside specifically for Centers of Cancer Nanotechnology Excellence, a program announcement with a specialized review dedicated to platform awards, and leveraging the general program announcement to support nanotechnology training centers. The Nanomedicine Initiative is also focused on the quantitative characterization of nanoscale components of the cell and the use of this knowledge to treat disease or repair tissue in a clinical setting. The new Brain Research through Advancing Innovative Neurotechnologies (BRAIN) Initiative is also expected to advance collaborative nanotechnology research and development efforts. The Administration announced the BRAIN Initiative in 2013 as part of a new Presidential focus aimed at revolutionizing the understanding of the human brain. The goal is to accelerate the development and application of innovative new technologies that build upon many scientific areas (e.g., neuroscience, genetics, physics, engineering, and nanoscience) to construct a dynamic picture of brain function that integrates neuronal and circuit activity for understanding behavior and diseases.

It is in early stages of planning, but as it matures, the Initiative will take advantage of new uses of nanotechnologies and create new ones.

Advances from the NCI Alliance expected for 2014 through 2015 include:

- Development and testing of a gold nanoparticle contrast agent in conjunction with clinical development of a Raman endoscope for enhanced colonoscopy.
- Nanoparticle formulations for rapid validation and treatment of potential therapeutic targets identified by genetic screens.
- Development of bioresponsive materials for imaging or detection of enzymatic activity *in vivo*, enabling noninvasive "optical" biopsies and diagnostic screens.

The end of 2015 will see the start of a five year third phase of the Alliance, supporting research and development centers, targeted platform awards, and interdisciplinary training centers.

Advances from the Nanomedicine Development Centers expected for 2014 through 2015 include:

- Development of alternative technologies based on nanopatterning and nanoarchitectures that stimulate the expansion of T cells for chronic lymphocytic leukemia.
- Development of bioreactor beads to manipulate the ratio of memory T Cells to effector T Cells in blood for treating leukemia (CLL).

In addition, NCI funding opportunities released in 2013 that are likely to exploit nanotechnology to extend basic biomedical research and advance disease diagnosis and treatment include:

- Development of Appropriate Pediatric Formulations and Pediatric Drug Delivery Systems FOAs.
- A continuation of the Image-Guided Drug Delivery in Cancer FOA.
- A continuation of the Research Answers to NCI's Provocative Questions FOA.

NHLBI continues to support the Programs of Excellence in Nanotechnology, with a focus on the development of nanotherapeutics, imaging agents, and *in vitro* diagnostic devices. Advances during the past year include development of a magnetic barcode assay for genetic detection of pathogens, permitting strain-specific, point-of-care identification of *M. tuberculosis*; nanoparticle-encapsulated siRNA used to silence inflammatory signaling; and large-scale single-step manufacture of drug-containing multifunctional high-density lipoprotein-like nanoparticles.

In summary, research activities throughout the entire "Nanotechnology Program Activities" sponsored by the institutes and centers of NIH will address future challenges in Nanotechnology Signature Initiatives (PCA 1), Foundational Research (PCA 2), and clinical translation/evaluation in support of NanotechnologyEnabled Applications, Devices, and Systems (PCA 3) commercialization.

NIST: NIST is developing optical picocalorimetry measurement instrumentation and methods needed to advance technologies ranging from semiconductors and data storage to biotechnology and optoelectronics. Measurement methods and instrumentation are also being developed for engineered nanomaterial metrology and characterization, including separations and surface composition. NIST is also developing the first SI-traceable nanoindentation instrument for measuring nanoscale mechanical properties of a broad range of materials; methods to elucidate chemical, mechanical, and electrical properties of molecular layers at the nanoscale; small-angle x-ray scattering methods for nondestructive, three-dimensional profiling of nanoscale patterns in semiconductor fabrication; and broadband, high-resolution, scanning probe microscopy for high-frequency characterization of nanoelectronic devices and materials.

NIST is supporting high-risk, leading-edge nanotechnology research that anticipates future measurement and standards needs of industry and science through the Innovations in Measurement Science (IMS) Program. Multidisciplinary IMS research teams are completing the development of measurement methods needed to characterize in three dimensions the essential nanoscale processes in photovoltaic materials and devices; bridging measurement length scales to advance graphene device technologies; and developing the motion metrology for micro- and nanomechanical systems needed to provide quality control during manufacturing and to increase device reliability.

NSF: NSF supports fundamental nanoscale science and engineering in and across all disciplines. NSF's nanotechnology research is supported primarily through grants to individuals, teams, and centers at U.S. academic institutions. The efforts in team and center projects have been particularly fruitful because nanoscale research and education are inherently interdisciplinary pursuits, often combining elements of materials science, engineering, chemistry, physics, and biology.

The NSF nanotechnology investment in 2013 supported about 5,000 active projects, over 30 research centers, and several infrastructure networks for device development, computation, and education. It impacted over 10,000 students and teachers. Approximately 150 small businesses were funded to perform research and product development in nanotechnology through the SBIR and STTR programs.

Three new directions planned for 2015 are two-dimensional nanoscale materials; chemical species and nanostructured composite materials; and nanoscale optics and photonics. [*See also the section below on coordinated activities in support of Objective 1.1.*]

USDA/FS: The Forest Service plans in 2015 to continue investing in fundamental research on cellulose nanomaterials. These fundamental research areas include interfaces between cellulose nanomaterials and other polymeric materials, new methods to separate cellulose nanocrystal from wood, basic morphology of cellulose nanocrystals, and others. The results of this research will provide foundations in understanding the properties of cellulose nanocrystals and in incorporating the renewable cellulose nanomaterial in polymeric composites, with potential application in automobiles, aerospace equipment, biomedical devices, and replacement for other plastics applications.

USDA/NIFA: USDA/NIFA nanotechnology efforts continue to advance the frontiers of interdisciplinary nanoscale science, engineering, and technology R&D endeavors to address a broad range of grand societal challenges and opportunities facing agriculture and food systems through various extramural research funding mechanisms. NIFA supports nanoscience research and development primarily through competitive and formula grants to individual researchers at land grant universities and other public and private research institutes.

Coordinated Activities with Other Agencies and Institutions Contributing to Objective 1.1

DOD/Army, universities, industry, and international organizations: The Army-sponsored Institute for Soldier Nanotechnologies (ISN) continues to maintain strong interactions with Army labs to investigate manipulating the intrinsic properties of matter at the nanoscale. This research effort has strong active research collaborations that have led to effective transitioning of ISN-developed materials, engineering capabilities, and computational expertise. These efforts have strengthened the intersection of scientific disciplines including fundamental research on metal-graphene interactions through atomistic modeling of dislocation behavior at the material interfaces to investigate the mechanical properties of metal-graphene composites. Alloy technology has been transitioned from the ISN to the Army Armament Research, Development, and Engineering Center (ARDEC), the Army Research Laboratory (ARL), and MIT startup Xtalic for scale-up, testing, and further development. Biomolecules and bio-nanomaterials research has been coordinated with the U.S. Army Medical Research and Materiel Command (USAMRMC), and potentially through the North Atlantic Treaty Organization (NATO) Mild Traumatic Brain Injury Working Group. Additional coordinated activities include ARL support for the polyfunctional scaffold effort through Defense Threat Reduction Agency (DTRA) projects and the Air Force Research Laboratory (AFRL). The DTRA project focuses on the development of self-cleaning coatings and other technologies to improve coatings, and the overall effort involves researchers from ARL, AFRL, and the Naval Research Laboratory (NRL), as well as the Army's Edgewood Chemical and Biological Center, in the evaluation of additives in Air Force topcoats.

DOD/DARPA and NIST: DARPA's MesoDynamic Architectures (Meso) program has exploited nonlinearities intrinsic to the mesoscale-nanoscale to produce frequency sources with performance and stability well beyond the state of the art for their size. This offers increased performance and new capabilities for military systems ranging from longer ranges for radios to maintaining radar performance on vibrating platforms. The Meso program is engaged with NIST for independent performance evaluation. Demonstrations using prototype devices can enable the Global Positioning System (GPS) tracking and longer-range radio transmissions at lower operating power. Typically, the performance of measurement devices is limited by deleterious effects such as thermal noise and vibration. Notable exceptions are atomic clocks, which operate very near their fundamental limits. Driving devices to their physical limits will open new application spaces critical to future DOD systems.

DOD/DARPA, NSF, NIST, and industry: The Semiconductor Technology Advanced Research Network (STARnet) program is jointly funded by U.S. member companies and DARPA. Management is provided by DARPA, the Air Force Office of Scientific Research (AFOSR), and the U.S. member companies with the Microelectronics Advanced Research Corporation (MARCO) administering day-to-day operations. DARPA routinely shares the results with NSF, NIST, and the other DOD component organizations. [*Also applies to Objective 1.2, where it is described in greater detail.*]

NIH/NIDCR and NIST: In 2013, the National Institute of Dental and Craniofacial Research (NIDCR) participated in an Interagency Agreement with NIST to develop clinical standards for resin composite-based dental restorative materials. These restorative materials contain bio-inert nanoparticles for enhanced mechanical properties and/or bioactive nanoparticles with potential antimicrobial properties. Examples include the development of a surface-enhanced Raman spectroscopy substrate for biofilm quantification using controlled nanotopography and new approaches to quantify demineralization using innovative data analysis procedures with nanometer resolution. NIST efforts in producing tooth-mimetic substrates have focused on early mineralization events involving development of restorative materials based on nanometer-size building blocks. NIDCR's role is to financially support these efforts and provide scientific and clinical advice to NIST through 2015.

NIST, DOE, and industry: Working with SBIR-supported companies, NIST and DOE are jointly developing synchrotron spectroscopy imaging instruments and methods to characterize the chemistry, structure, and electronic state of engineered nanomaterials (ENMs). These instruments are being transferred to the new DOE synchrotron facility NSLS-II in 2014 and are expected to be fully operational in 2015. NSLS-II was designed specifically for nanoscale imaging and thus will provide enhanced capability of NIST's instruments for both spatial and depth resolution.

NIST, USDA, universities, and industry: NIST, the USDA Forest Product Laboratory, Purdue University, the University of Maine, and the Technical Association for the Pulp and Paper Industry (TAPPI, a non-profit organization supporting the American paper and pulp industry) published *Production and Applications of Cellulose Nanomaterials*, a book highlighting the importance of cellulose nanomaterials and the wide range of research being conducted worldwide to advance this technology.

NIST and industry: NIST is collaborating with industry on the development of standards for scanning microwave microscopy for applications in life sciences, next-generation electronics, and data storage technology.

NIST, universities, and industry: In December 2013, NIST selected a consortium led by Northwestern University to establish a new NIST-sponsored center of excellence for advanced materials research. The new Center for Hierarchical Materials Design (CHiMaD) will be funded in part by a $25 million award from NIST over five years. This partnership is an important component of NIST's programs in support of the Administration's Materials Genome Initiative. NIST will work closely with the university and industry members of the CHiMaD consortium to accelerate the design, discovery, development, and deployment of advanced materials. Nanotechnology and nanomaterials research will play an important role in this new center.

NSF and DOD/AFOSR: Both agencies collaborate in the Two-Dimensional Atomic-layer Research and Engineering (2-DARE) program, which addresses fundamental challenges in the creation of 2-D nanoscale materials and their hybrids, scalable manufacturing strategies, characterization tools and methods, and novel devices. These may lead, among other things, to solar cells that exploit the unique properties of stacked, heterogeneous layers.

NSF, EPA, and USDA/NIFA: NSF, EPA, and USDA collaborated in a joint grantees conference in December 2013, facilitating portfolio evaluation and identification of future research directions.

Objective 1.2 – Identify and Support Nanoscale Science and Technology Research Driven by National Priorities and Informed by Active Engagement with Stakeholders

Individual Agency Contributions to Objective 1.2

DOD/Army: ARL and the Institute for Soldier Nanotechnologies collaborated in a malaria vaccine study using a murine model. A new ISN-developed nanoparticle, called an inter-bilayer-crosslinked multi-lamellar vesicle, provided a tenfold increase in immune response with 100 times less vaccine. Moreover, the immune response remained essentially unchanged for over five months. Along a different direction is the ongoing development of "rapid reconstitution packages" of lyophilized drugs to enable greatly simplified logistics, dramatically prolonged drug shelf life, and increased ease of administration in the field.

DOD/DARPA: A competitive electronics industry is a national priority to maintain commercial technology leadership and ensure new defense capabilities. DARPA is identifying promising new approaches for using nanotechnologies to drive advances in new devices to replace transistors or to augment the performance of conventional transistor integrated circuits. See the DOD STARnet and Meso program descriptions under Collaborative Activities for Objective 1.1. The STARnet program is advancing fundamental research for creating highly functional circuits using these devices and understanding their inherent limitations. In this case, the program is partnering directly with several industrial stakeholders including equipment supplier Applied Materials, semiconductor integrated device manufacturers Intel and Micron, foundries IBM and Global Foundries, along with defense contractors Raytheon and United Technologies. The STARnet program funds research directly at U.S. universities.

Many defense-critical applications require exceptionally precise time and frequency standards enabled only by atomic clocks. GPS and the Internet are two key examples. DARPA's Enabling Quantum Technologies program is building upon established control and readout techniques from atomic physics to develop a suite of measurement tools that will find broad application across the DOD, particularly in the areas of biological imaging, inertial navigation, and robust global positioning systems. Recent progress under this program includes the demonstration of the world's most precise atomic clocks that exhibit stability greater than two parts per quintillion (one followed by 18 zeros).

Defense applications such as geolocation, navigation, communication, coherent imaging, and radar depend on the generation and transmission of stable, agile electromagnetic radiation. Improved radiation sources—for example, lower-noise microwaves or higher-flux x-rays—could enhance existing capabilities and enable entirely new technologies. The Basic Photon Sciences program is investigating ways to exploit the exquisite stability and control achieved in ultrafast lasers to enable novel radiation sources that span the electromagnetic spectrum with improved performance while reducing size, weight, and power. It is envisioned that such devices, once demonstrated by DARPA, will find broad application across the DOD, particularly in the areas of secure communications, electronic warfare and countermeasures, coherent radar, nondestructive evaluation, biological imaging, and materials characterization. Recent key achievements in this program include the demonstration of up to 20 decibel reduction in the noise floor of photonic microwave generation compared to state-of-the-art lab-based sources and the first demonstration of 3-D full-field x-ray imaging with 5.5 nanometer spatial resolution and about 1 femtosecond temporal resolution.

DOD/Navy: The Navy nanomanufacturing program is developing nanoscience-based, high-speed continuous fabrication of full-function hybrid flexible electronic systems. The challenge is to form a family of devices (sensors, logic circuits, radio frequency identification—RFID, displays, power, etc.) as a multifunctional system-on-film. The program is developing appropriate nanomaterials (oxides, CNTs, silver, etc.) that will be used in the building of multilayered devices; appropriate nanoscale processes (printing, imprinting, etc.) for creating the circuits in an optimal sequence; design and construction of the circuits for selected systems of interest optimized for performance and production; and strategies for continuous, rollto-roll nanomanufacturing of these complex electronic systems. Flexible electronics have the potential to be lightweight and low-cost, both features beneficial to the military. The program is targeting an efficient, conformal brain-machine interface (BMI) electronic system, which will replace a current discrete electrode-based BMI.

DOE: DOE-SC continues to support the Joint Center for Artificial Photosynthesis (JCAP), an Energy Innovation Hub aimed at developing revolutionary methods to generate fuels directly from sunlight. In 2013 DOE-SC initiated the Joint Center for Energy Storage Research (JCESR), another Energy Innovation Hub focused on batteries and energy storage. A significant fraction of the effort in both hubs involves nanotechnology. By fostering unique, cross-disciplinary collaborations, these hubs will help advance highly promising areas of energy science and engineering from the early stage of research to the point where the technology can be handed off to the private sector.

The Advanced Research Projects Agency-Energy (ARPA-E) supports a broad range of early-stage, potentially disruptive energy technology projects that involve both applications of current nanotechnology and applied research and development in new areas of nanotechnology. A major focus is on the challenging translation of nanotechnology from the laboratory to commercial viability.

The SunShot Initiative in the DOE Office of Energy Efficiency and Renewable Energy (DOE-EERE) supports a variety of projects that use nanotechnology to drive down the cost of solar power installations. Approaches to reach the SunShot $1 per watt target using nanotechnology include quantum dot solar cells, nanostructures for bandgap engineering, monolayers to adjust a material's work function or passivate surfaces, and nanostructured materials for plasmonics and antireflective coatings.

The solid-state lighting (SSL) program in DOE-EERE supports research and development of promising SSL technologies in areas such as emitter materials research at the nanoscale level to reduce current droop and thermal sensitivity in light-emitting diodes (LEDs) and light extraction approaches for organic LEDs (OLEDs). Manufacturing R&D projects accelerate SSL technology adoption through manufacturing improvements that reduce costs and enhance quality of SSL devices.

The Hydrogen and Fuel Cell Technologies (HFCT) program in DOE-EERE conducts nanomaterials research and development for fuel cell, hydrogen production, and hydrogen storage applications. These efforts, such as those in catalyst R&D, will lead to lower costs and improved performance in hydrogen and fuel cell technologies and will help to enable widespread commercialization of these technologies.

DOT/FHWA: The Federal Highway Administration is developing innovative materials and coatings that can provide significant improvements in the durability, performance, and resiliency of transportation infrastructure. Nanomaterials efforts are focused on improving the behavior and durability of cementitious materials through modification and enhancement of interface behavior and expanding the fundamental understanding of cement hydration kinetics. Multifunctional infrastructure materials can improve structural health monitoring, self-repair capability, and the ability to detect and mitigate corrosion. Advancements in nanoscale sensors will provide opportunities for networked infrastructure and connectivity with vehicles to provide a safer, more efficient, longer-lasting transportation system.

NIH: NCI has engaged the external stakeholder community through a strategic workshop, the Request for Information on Needs and Directions in Cancer Nanotechnology, and a web-based forum entitled *Dialogue on Nanotechnology in Cancer*, which has over 100 registered users. The community response to these efforts is being used to identify the pressing needs and promising opportunities for cancer nanotechnology and to inform the shape of NCI's future investment in nanotechnology. Nanotechnologyspecific responses were also received in response to the Request for Information (RFI), "Regarding Opportunities for Cancer-Relevant Innovative Technologies with Transformative Potential." NCI also engages with the patient advocate and pharmaceutical/biotechnology industry through the Translation of Nanotechnology in Cancer (TONIC) consortium. The collective stakeholder inputs to these efforts were integrated into the successful Alliance Program renewal effort and will be key to identifying programmatic priorities as Phase III of the Alliance begins at the end of 2015. [*See Objective 2.2 for more details.*]

NIOSH: NIOSH continues to provide broad support to advancing nanotechnology research through active partnership with the private sector and through the NNI Signature Initiatives. Within this effort, NIOSH has several activities that are focused directly on safe and responsible development of the technology that will facilitate speed to market and competitive advantage.

NSF: NSF organized the Arizona State University Center for Nanotechnology in Society (CNS-ASU) first annual Winter School on the Anticipatory Governance of Emerging Technologies focused on statistical data on societal dimensions of nanotechnology, socio-technical integration research methods, and future scenario development.

USDA/ARS: Nanotechnology investments include research in the following areas:

- Processing technologies to produce healthy, value-added foods from specialty crops, including applications of methylcellulose and chitosan nanoparticles in edible films.
- Imaging technology for food safety and security, including use of Raman scattering with silver nanorods or use of nanoscale peptides and DNA aptamers to detect food-borne pathogens and toxins.
- Novel methods for manufacturing of bioproducts from agricultural feedstocks, including blow-spinning and electro-spinning of biopolymers to produce nanocomposites, nanofoams, nanofibers, microemulsions, and hydrogels.
- Environmentally friendly processes and new applications for animal hides and leather, including collagen-based nanofibers for high-efficiency, biodegradable air filters.
- Development of new bioactive and bio-based products from plant cell wall polysaccharides in sugar beet pulp, citrus peel, and other processing residues, including polysaccharide-based nanoparticles for controlled release of bioactive food ingredients and for active packaging applications.
- Development of nanoscale coatings to make cotton flame-resistant, and depositing enzymes/peptides on cotton nanocrystals for applications in biosensors, antimicrobials, and bioremediation.

USDA/NIFA: The NIFA nanotechnology program supports foundational sciences and innovative ideas to develop nanotechnology-enabled solutions for a range of national priorities including global food security through improving productivity, quality, and biodiversity; adaptation and mitigation of agricultural production systems to climate changes; improvement of the nutritional quality of food and feed; early detection and effective intervention technologies for ensuring food safety and biosecurity; more effective therapies that significantly impact animal health and wellness; development of the biology-based economy; and increased protection for natural resources, the environment, and agricultural ecosystems.

Coordinated Activities with Other Agencies and Institutions Contributing to Objective 1.2

DOE and NSF: NSF and DOE co-sponsor the Quantum Energy and Sustainable Solar Technologies (QESST) Engineering Research Center (ERC), which conducts photovoltaic (PV) research. The DOE Hydrogen and Fuel Cell Technologies (HFCT) program activities are coordinated primarily with DOE-SC and NSF, working to understand the fundamental science behind relevant nanomaterials. The findings from the fundamental work are then applied to the HFCT R&D efforts to further develop and then to commercialize nanoscale technologies in hydrogen and fuel cells.

NASA, NRO, and DOD: In 2013, NASA and the NRO participated in a technical interchange meeting to identify areas of mutual interest and explore opportunities for further collaboration in nanotechnology R&D. NASA continues to work with AFOSR in carbon nanotube and boron nitride nanotube structural materials.

NSF, NIH, EPA, DOD, NASA, and USDA/NIFA: "Nanotechnology Research Directions for Societal Needs in 2020" (see wtec.org/nano2), the report of a study led by NSF with input from academic, industry, and government experts from over 35 countries, addresses the progress and impact of nanotechnology since 2000 as well as the vision and research directions for nanotechnology by 2020. Further, convergence of nanotechnology with other technologies and areas of application was analyzed in the NSF-led study and 2013 report created in collaboration with NIH, EPA, DOD, NASA and USDA/NIFA, *Convergence of Knowledge, Technology, and Society* (see: wtec.org/NBIC2-Report).

NSF, NIST, and industry: The Semiconductor Research Corporation (SRC) works with NIST and NSF to address the confluence of two key national priorities that are also associated with long-standing U.S. Government interagency research initiatives: (1) nanotechnology/NNI and (2) information technology/NITRD (Networking and Information Technology Research and Development).

Objective 1.3 – Assess the Performance of the U.S. Nanotechnology R&D Program

Individual Agency Contributions to Objective 1.3

DOD/Air Force: Exploration of the applications impact of emerging nanomaterials work is of principal importance in the AFRL nanotechnology research portfolio. Nanomaterials and nanodevice research is informed by advances in nanomanufacturing to determine where application goals can be met. Nano-optical materials research is focused on quantitative improvements to filter technology and optical detector technology by exploring quantitative improvements in performance through novel zero-index nanomaterials and nanodevice enhancements of infrared detectors including increased sensing functionality through transformation optics in a nano-patterned system. In addition, an investigation of nanomaterials for improving the performance and function of additive manufacturing systems has been conducted. Evaluations of technical progress were done by comparison with conventional materials approaches to similar technical challenges. [*This is also relevant to Objective 4.1 through the required evaluation of new process hazards.*]

NIH: A thorough performance assessment was a critical component of the successful renewal of the NCI Alliance for Nanotechnology in Cancer (Phase III to start at the end of 2015). As noted in Objective 1.2, NCI combined the results of a pilot analysis of Alliance network connectivity and productivity with NIH portfolio analysis, bibliometric analysis of Alliance publications, input from an external evaluation committee and other experts in the field, and responses to relevant questions from the Cancer Nanotechnology RFI to assess the multidisciplinarity, translational potential, and productivity of the Alliance program. This evaluation effort addressed other important issues including the appropriateness of the balance between basic and translational research in the program, the effectiveness of training efforts in the program, the extent to which the program advances adherence to standards by researchers in the field, and the value of the program's engagement with the clinical and industrial communities. The results of this assessment inform the structure and strategy of NCI's continuing nanotechnology investment.

NSF: The CNS at ASU has evaluation activities for research performance and nanotechnology outcomes. The CNS node at Georgia Tech evaluates trends in publication, patents, and commercialization through data mining, interviews, and other methods as part of the "Research and Innovation Systems Analysis" group. The group's main goal is to characterize the technical scope and dynamics of the nanoscale science and engineering enterprise and the linkages between it and a variety of public values and outcomes.

USDA/NIFA: USDA/NIFA continues its evaluation of the performance and progress of NIFA's nanotechnology R&D projects through its annual grantees meeting. Given the tight travel budget, the 2014 grantees meeting will be conducted in video conferences.

USPTO: The U.S. Patent and Trademark Office contributes a variety of nanotechnology-related patent data, which is used as a benchmark to analyze nanotechnology development and for trend analysis of nanotechnology patenting activity in the United States and globally. The data shows that since the start of the NNI, the percent of nanotechnology-related U.S. patents assigned or owned by universities have slightly increased and the percent of nanotechnology-related U.S. patents with government interest statements has almost doubled. Both of these increases are likely attributable to the increased Federal R&D funding from the NNI. Regarding international nanotechnology-related patenting activity, U.S. inventors continue to lead in the total number of nanotechnology-related patent publications globally. However, that lead has been shrinking. U.S. inventors more significantly lead in the total number of nanotechnologyrelated patent publications in three or more countries, indicating an aggressive pursuit of international intellectual property protection.

Coordinated Activities with Other Agencies and Institutions Contributing to Objective 1.3

DOD/Air Force, universities, and industry: The teams executing the Air Force efforts towards this objective (listed above under "individual agency contributions") include academic-industrial partnerships fostered by interaction through AFRL and industry partnerships. The specific information has been used both for comparison with technical goals in these areas and to evaluate the current overall nanomaterials engineering capability.

NNI agencies, NNCO, National Academies, and the President's Council of Advisors on Science and Technology (PCAST): As discussed in Section 5 of this report, the NNI participating agencies and NNCO contribute background information and other resources, as appropriate, to the periodic reviews of the NNI by the National Academies and PCAST. These reviews include assessment of the performance of the nanotechnology research supported by the NNI agencies.

Objective 1.4 – Advance a Portfolio of Nanotechnology Signature Initiatives (NSIS) That Are Each Supported by Three or More NNI Agencies and Address Significant National Priorities[17]

Nanotechnology for Solar Energy Collection and Conversion

Individual Agency Contributions to This NSI

DOD/Air Force: The Air Force Materials and Manufacturing Directorate contributes to this NSI by continuing to develop nanostructured organic photovoltaic devices based on single-layer and tandem bulk heterojunction (BHJ)-type photovoltaic devices. Current activities are focused in three areas: (1) developing approaches to fabricate these devices on nontraditional substrates to enable structural power applications (such as solar aircraft wings), (2) designing reproducible recombination layers for tandem (multilayer) devices, and (3) developing direct-write approaches for on-demand, low-cost, scalable production of these devices. These devices are expected to have the greatest impact for applications requiring very high specific power (W/g) such as unmanned air vehicles and high-altitude airships.

DOE: The Solar Energy Technologies Program (SETP) in the DOE Office of Energy Efficiency and Renewable Energy supports a variety of efforts related to nanotechnology research and development in this NSI at the intersection of scientific disciplines. For example, the Bridging Research Interactions through collaborative Development Grants in Energy (BRIDGE) Funding Opportunity Announcement (FOA) was released to support collaborative efforts between scientists with photovoltaic experience and those with expertise utilizing the national scientific user facilities. The BRIDGE FOA was executed as a collaborative effort between DOE-SETP and DOE-SC. A variety of projects funded under this FOA are related to nanotechnology, such as those focused on grain boundaries and nanoscale arrays of photoactive materials.

The DOE Office of Science supports a wide array of fundamental research related to solar energy, including a number of Energy Frontier Research Centers (EFRCs) and the Joint Center for Artificial Photosynthesis (JCAP).

NIST: NIST is developing measurement methods needed to characterize, in three-dimensions, the essential nanoscale processes in photovoltaic materials and devices, and is expanding efforts into methods to be applied during manufacturing. In support of this NSI, NIST has demonstrated a new solar simulator system with the ability to match arbitrary solar spectral conditions while maintaining spot size below 2000 nanometers, enabling optimization of advanced photovoltaic materials such as quantum dots, polycrystalline semiconductors, and multijunction stacks.

NSF: NSF supports basic research and interdisciplinary efforts by groups of primarily academic researchers to address the scientific challenges of highly efficient harvesting, conversion, and storage of solar energy— projects in the Mathematical and Physical Sciences (MPS), Engineering (ENG), Computer & Information Science & Engineering (CISE), and Biological Sciences (BIO) Directorates. A primary emphasis is on the education, at all levels, of a highly trained scientific and technical workforce. NSF activities that address research topics generally related to this NSI include Engineering Research Centers, such as the QESST ERC,[18] and Emerging Frontiers in Research and Innovation (EFRI).

The Sustainable Energy Pathways (SEP) program is part of the NSF-wide initiative on Science, Engineering, and Education for Sustainability (SEES). The SEP solicitation calls for innovative, interdisciplinary basic research in science, engineering, and education by teams of researchers. One example of an SEP program that addresses the goals of this NSI is the Development of Economically Viable, Highly Efficient Organic Photovoltaic Solar Cells Program, which is administered through the University of Chicago, Northwestern University, and the University of California, Los Angeles. This program is aimed at achieving low-cost solar energy conversion to electricity through the development of organic photovoltaic (OPV) solar cells, through evaluating the sustainability of OPV technology, and through addressing the environmental and economic aspects of OPV technology.

The MPS and ENG Directorates are responsible for over 80% of the NSF investment. Examples of investments relating to this NSI include research on photovoltaic energy conversion theory, modeling, surface chemistry, materials synthesis, device physics, engineering and characterization, and nanotechnology for power electronics.

USDA/NIFA: The NIFA Agriculture and Food Research Initiative (AFRI) nanotechnology program supports a broad range of value-added uses of agricultural bio-nanomaterials and products to strengthen the national economy while seeking innovative technologies to provide new energy sources. A successful research example is the development of direct conversion of solar energy to electricity using nanoscale protein units responsible for plant photosynthesis from leafy vegetables and other plants. Another example is the development of a prototype solar battery developed by a multidisciplinary research team at Vanderbilt University that utilizes a nanoscale photosystem I (PSI) multilayer to generate enough power to operate a handheld calculator.

Coordinated Activities with Other Agencies and Other Institutions Contributing to This NSI

DOD/Air Force, NNI agencies, and universities: The Materials and Manufacturing Directorate work is coordinated with other DOD and Government agencies through both the Reliance 21 Communities of Interest (Advanced Materials and Energy & Power) as well as through the Interagency Advanced Power Group. AFRL collaborates with the Georgia Institute of Technology, Northwestern University, and the University of Chicago.

NNI agencies: Agencies participating in the NSI collaboratively organized a symposium on Nanotechnology for Solar Energy Collection and Conversion at Nanotech 2013[19] (May 2013, in Washington, DC) and participated in meetings of the NSI group to exchange information and coordinate nanotechnology efforts in this field.

NSF and DOE: In addition to continuing the joint funding NSF-DOE activities in the Solar Energy NSI, NSF and DOE are co-sponsoring the Engineering Research Center for Quantum Energy and Sustainable Solar Technologies (QESST ERC), another research center that involves industry collaboration. This center has been jointly funded by the DOE's Solar Energy Technologies Program and NSF since 2011. QESST research focuses on developing a fundamental understanding of electronic excitations, relaxation, and transport in nanostructured materials, leading to materials with electric-charge collection efficiencies in excess of the single-junction solar cell limit. Under development in this research center are sophisticated new III-V quantum-scale semiconductor materials and solar designs with projected efficiencies in excess of 50%, and new photovoltaic energy concepts that will enhance planetary and deep space exploration. This center also works with companies to advance technologies toward commercialization.

NSF, DOE, and NIST: The annual "Photovoltaic Specialist Conference," which is co-sponsored by DOE, NSF, and NIST, continued to be held in each of the years following the articulation of this NSI (2010, 2011, 2012, and 2013). This conference also has included a workshop titled "PV Velocity Forum: Accelerating the PV Economy," that was jointly planned and implemented by DOE and NIST.

NSF, industry, and universities: Collaborative research centers have facilitated ongoing partnerships between participating agencies, industry, and academia, thereby helping to strategically focus R&D efforts and efficiently implement R&D advances. For example, the "Silicon Solar Consortium" Research Center is an industry/university collaborative research center that involves joint NSF and industry funding. This center performs research on crystalline silicon crystal growth, impurities and defects, photovoltaic device processing, process modeling and simulation, and module issues. This center has created a multi-university, multi-company culture addressing the science and technology issues that the international photovoltaic silicon materials industry must solve in order to meet the future needs of advanced silicon solar cell manufacturing, while educating graduate students with photovoltaic materials/devices expertise to meet future workforce needs. As a collaborative agency/industry enterprise, this center conducts the critical research related to improving the efficiency of PV devices while addressing the manufacturing issues related to realizing the commercialization of these devices.

Sustainable Nanomanufacturing: Creating the Industries of the Future

Individual Agency Contributions to This NSI

DOD/Air Force: The Materials and Manufacturing Directorate is conducting numerous research efforts toward nanoelectronics, with efforts focused on enhanced electron and phonon transport, carbon nanotubes, 2-D and 3-D electron gas structures, oxide nanostructures, thin film processes, and predictive modeling of materials performance for applications in surveillance and electronic warfare. Significant efforts throughout the Directorate are focused on the development of metallic, organic, and carbon nanotube inks for a wide range of nanoelectronic devices, including flexible, inherently rugged electronic devices.

DOD/Army: The Army Research Laboratory has successfully engineered a new method of producing stable nanocrystalline powders with sufficient stability to resist microstructural changes during the high-temperature processing needed for full-density consolidation. Availability of advanced nanocrystalline alloys is critical to the paradigm shift mandated for future Army infantry, weapons, and vehicle performance. Nanocrystalline microstructures thermodynamically are subject to coarsening at room temperature. Therefore, the processing and applications of these materials will be limited by the ability to create stabilized microstructures. Current users exploit state-of-the-art atomic-level characterization techniques to identify and determine the electronic contributions responsible for reducing and minimizing the interfacial energy of high-temperature nanocrystalline alloys that are relevant to Army applications. Through this process researchers have been able to bulk consolidate and/or produce near-net-shaped parts, retaining nanocrystalline microstructure, potentially suitable for various Army applications. These include aluminum alloys for light-weighting ballistic applications, nanostructured steels for ultrahigh strength at high temperatures, and nanostructured materials for penetrator applications.

NASA: NASA continues to work with industry, universities, and other Federal agencies in the development of ultralightweight, high-strength composites based on carbon nanotube reinforcements and their application in aerospace components. This project, supported by the Space Technology Mission Directorate's Game Changing Development Program, will design, build, and flight-test a carbon nanotube composite reinforced pressure vessel in 2015. In 2013, NASA researchers, under a Center Innovation Fund project, have been exploring the use of halogen intercalation chemistries to improve the electrical conductivity of carbon nanotube yarns and fibers for use in lightweight data and power cables. This work is continuing in 2014 with support from the Game Changing Development Program.

NIOSH: NIOSH continues to provide broad support to advancing nanotechnology research through the NSIs. Within this effort, NIOSH has several activities that are focused directly on safe and responsible development of the technology. In 2015, NIOSH will continue to develop and disseminate case studies that demonstrate the utility of applying Prevention through Design (PtD) principles to nanomanufacturing. Elements of the PtD program include identification of physicochemical properties of nanomaterials that can be modified to decrease adverse biological responses, and the development of EHS practices for incorporation into the texts and curricula of U.S. engineering schools. Partnerships with private nanomanufacturing companies will be expanded in 2015 to include extension of effective risk management practices into elements of the Administration's Advanced Manufacturing priority area. Many nanomaterial-enabled applications and products will realize full commercialization through advanced manufacturing technologies, and NIOSH will extend and reapply knowledge gained from nanomanufacturing into this area. [*Funding for this activity is reported by NIOSH under PCA 5; also relevant to Objective 4.4.*]

NIST: NIST is developing high-throughput measurement methods to characterize the nanoscale morphology of carbon nanocomposites. These microwave-based measurement methods will be used in manufacturing for quality assurance and control. As part of this program, new high-resolution techniques are also being developed to provide insight into the processing-structure-properties relationships in nanocomposites. In particular, a novel method for actuating atomic force microscope cantilevers is yielding mechanical property information at a length scale of 30 nanometers, helping elucidate the role of nanoscale reinforcements in composite behavior.

NSF: The NSF Nanomanufacturing Program and other core programs fund fundamental research projects and have contributions in carbon-based nanomaterials, optical metamaterials, and cellulosic nanomaterials. A program solicitation on Scalable Nanomanufacturing has had two competitions, in 2013 and 2014; another one is planned for 2015.

USDA/NIFA: NIFA's current focus in nanomanufacturing is on nanobiomaterials derived from various forest, crop, and by-products biomass. There are ongoing research efforts in synthesis of carbon-based nanomaterials; development of cost-effective production methods; functionalization and characterization of nanomaterials; and exploring applications such as piezoelectric, renewable nanocomposite polymers, flexible displays, food packaging, barrier films, windows, and others.

Coordinated Activities with Other Agencies and Other Institutions Contributing to This NSI

DOD/Army, NNI agencies, and universities: Army efforts are highly embedded with universities through collaborations with several national universities including George Mason University, MIT, Penn State, University of North Carolina, North Carolina State University, Drexel University, and Texas A&M. Other partners include Los Alamos National Laboratory, the ARL Sensors and Electron Devices Directorate, and the Army Armor Research, Development, and Engineering Center. There are collaborations with the Center for Computational Materials Design, an NSF-supported Industry/University Cooperative Research Center, on various aspects of phase formation and grain growth. The ARL internal mission program was instrumental in developing an ARL/ARO SBIR project focused on the processing and property interrogation of high-strength nanostructured aluminum alloys. This program directly impacts the Army light-weighting capability by seeking to develop alloys that exceed the expected physical properties of current aluminum alloys. This would enable weight reduction of 65% in alloys with ultrahigh toughness.

NIST, NASA, universities, and industry: NIST is working with samples from NASA, a major aerospace company, and MIT to validate new measurement methods that enable a very detailed analysis of the distribution of nanomaterials in polymers—information that is critical to turning these promising high-performance composites into flight-qualified components.

NNI agencies: Agencies participating in the NSI collaboratively organized a symposium on Sustainable Nanomanufacturing at Nanotech 2013[20] (May 2013 in Washington DC) and participated in meetings of the NSI group to exchange information and coordinate nanotechnology efforts in this field.

NRO, DOD, industry: NRO is working with DOD and several companies to leverage Defense Production Act Title III funds to help develop and expand U.S. industrial capabilities for nanomanufacturing of bulk carbon nanotube sheets, yarns, and tapes, as well as compression-molded carbon nanotube composite parts. [*See additional details below under activities in support of Objective 2.2.*]

USDA/FS and universities: In 2013, the University of Maine commissioned a Forest Service-funded 1 ton per day pilot-scale production facility for making cellulose nanofibrils using mechanical methods. This pilot facility is the second of the two Forest Service-funded pilot facilities for cellulose nanomaterials research and development. The first of these, located at the Forest Service's Forest Products Laboratory in Madison, WI, produces both cellulose nanocrystals and nanofibrils using a different process from the Maine plant. It was commissioned in 2012.

The Forest Service continues to collaborate with a network of universities in cellulose nanomaterials (CN) research. This Forest Service-academia collaboration was established with a one-time investment in 2009. In the 2013 progress review held at Oak Ridge National Laboratory, researchers presented their results for cellulose nonmaterial research in the areas of CN functionalization, aerogels from CN, composites with improved wet strength and liquid barrier properties, predictive modeling, aqueous aerospace composites from CN, and online measurement of CN in a manufacturing process.

Nanoelectronics for 2020 and Beyond

Individual Agency Contributions to This NSI

DOD/Air Force: The Materials and Manufacturing Directorate is exploring nanoelectronic materials and their synthesis processes to improve device performance and reliability and to develop technologies that are less dependent on foreign foundries. Currently, most electronic materials are manufactured abroad, which places the United States at a strategic disadvantage for next-generation electronic materials. Alternative processes for electronics manufacturing are being explored to allow for a cost-effective and rapid progress in advanced nanoelectronic materials, which can include 0-D (e.g., quantum dots), 1-D (e.g., nanotubes and nanowires), and 2-D (e.g., monolayer) approaches. These materials and processes are of strategic importance for both DOD applications (sensors, communications, navigation, etc.) and for commercial products in telecommunications, computation, medicine, and other fields. AFRL is actively researching these nanoscale electronic materials and their processing technologies as well as developing strategic partnerships with U.S. academic research centers and industry for both research advancements and applications development.

DOD/Army: ARL developed a breakthrough materials synthesis strategy that results in a fifty-fold increase in the crystal size of hexagonal boron nitride (hBN) 2-D nanomaterials. Optimized hBN is needed to serve as the insulating interface for graphene nanoelectronics, thus this work is a critical step in nanoelectronics development.

NIST: NIST is making significant advances in nanoelectronics by developing measurements and fabrication approaches for high-performance devices based on semiconductor nanowires and two-dimensional thin film device platforms for electronic and nanophotonic applications; advancing processes to integrate molecular components into robust electronic devices with metal or silicon electrodes; developing new approaches to characterize topological insulators; characterizing the energy barriers at materials interfaces critical for future ultra-low-powered electronics such as tunnel FETs; creating new nanoscale methods to measure and understand spin transport, dynamics, and related magnetic interactions in electronic and data storage devices; developing novel ways to measure and manipulate photons in nanophotonic applications; and coupling quantum optical and nanomechanical systems in silicon to create innovative devices for metrology and quantum communications applications. In addition, NIST is making major advances in carbon-based nanoelectronics, having developed a rapid and economical method to produce high-purity samples of carbon nanotubes along with techniques to monitor their purity, and by elucidating critical properties of graphene and graphene-based devices. Critical advances in materials growth and nanofabrication have been made at NIST that open the way for the use of silicon-based nanoelectronics in quantum information technology. The NanoFab, a shared resource in NIST's Center for Nanoscale Technology (CNST) user facility, is providing support to this NSI through access to a comprehensive commercial-state-of-the-art tools set for nanoelectronics fabrication.

NSF: The NSF investment has a primary focus on upstream, exploratory research in the following areas: new alternative "state variables" for logic and memory components and suitable computer architectures, computer-based research of new devices, integration of nanoelectronic and nanophotonic components into new systems, and new quantum information system components and systems. Examples of core programs covering this thrust are the Electronic and Photonic Materials Program within MPS's Division of Materials

Research and the Electronics, Photonics and Magnetic Devices Program within ENG's Division of Electrical, Communications and Cyber Systems. Examples of center activities include the Materials Research Science and Engineering Centers (MRSECs), the Science and Technology Centers (STCs), the Engineering Research Centers, and the Nanoscale Science and Engineering Centers (NSECs). Two examples are the STC on Quantum Materials and Devices at Harvard University and the Center for Quantum and Spin Phenomena in Nanomagnetic Structures at the University of Nebraska, Lincoln. NSF also supports the National Nanotechnology Infrastructure Network (NNIN), whose 14 universities provide capabilities that include nanoelectronics.

Coordinated Activities with Other Agencies and Other Institutions Contributing to This NSI

DOD/Army and universities: Coordinated activities include Nanyang Technological University (Singapore) with research goals focusing on 2-D nanomaterial synthesis, including hexagonal boron nitride (hBN).

NSF, NIST, industry, and universities: Via the Nanoelectronics Research Initiative (NRI), NNI agencies and industry support 12 interdisciplinary research teams at academic institutions across the Nation.

NNI agencies: Agencies participating in the NSI collaboratively organized a symposium on Nanoelectronics at Nanotech 2013[21] (May 2013 in Washington, DC) and participated in meetings of the NSI group to exchange information and coordinate nanotechnology efforts in this field.

Nanotechnology Knowledge Infrastructure (NKI): Enabling National Leadership in Sustainable Design

Individual Agency Contributions to This NSI

NIH: NCI supports the caNanoLab, a web-based portal and data repository that allows researchers to submit and retrieve information on nanoparticles including composition; function (e.g., therapeutic, targeting, diagnostic imaging); physical characteristics (e.g., size, molecular weight); and *in vitro* (e.g., cytotoxicity, immunotoxicity) experimental characterization, along with information on the protocols used for these characterizations and links to any related publications. NCI is working with journal editors to promote the use of the portal by authors who, concurrent with publication, would be encouraged to submit information to caNanoLab to continue to develop the portal into a robust outlet for storage of characterized nanomaterials data and detailed data sharing. Currently, the caNanoLab website averages 80–120 unique visitors per month. Software from caNanoLab is available for download and installation at local institutions. It is open source, and the code is being deposited to the National Cancer Informatics Program (NCIP) channel in the GitHub code repository to support open, community-led development. Going forward, the caNanoLab team aims to emphasize policies and resources that promote and incentivize standards-based data capture directly by the data producers. [*Also relevant to Objective 4.1.*]

NIOSH: NIOSH remains focused on developing an effective nanoinformatics tool for EHS practitioners. Specific contributions include ongoing support for the Nanoparticle Information Library and making available real-life data on current and emerging nanotechnology practices in the workplace. This latter investment will include the collection of data on workplace exposures; experimental evidence and modeling of toxic effects from categories of nanomaterials; knowledge and data on efficacy of controls; recommended exposure limits; and tools and guidance to support sustainable nanotechnology. [*Funding for this activity is reported by NIOSH under PCA 5; also relevant to Objective 4.4.*]

NIST: NIST is developing a series of standard procedures to validate theoretical models and computational algorithms used in the prediction of a wide variety of physicochemical properties of nanomaterials. NIST efforts include validated models for direct comparison with nanomechanical property measurements of nanowires with the goal of producing predictive models; software to compute interactions between magnetic spins for visualization of nanoscale domain structures; and a database of computationally generated electrical properties and structures of metal nanoparticles. The NKI provides essential opportunities for interagency collaboration in support of these activities.

NSF: NSF continues its participation in the NKI through investments in nanoscale modeling and simulation, database networking and computer user facility (Network for Computational Nanotechnology, or in brief, nanoHUB), and Cyber-Enabled Discovery and Innovation initiatives within its Cyberinfrastructure Framework for the 21st Century Science and Engineering activities such as Software Infrastructure for Sustained Innovation. NSF will also contribute to the foundations of this NSI by means of specific databases for nanoscale materials and processes, transformative thinking about models for linkage of properties and behaviors at different scales, extension of computational and statistical techniques to support development and use of the nanotechnology cyber toolbox to accelerate discovery and manufacturing of nanomaterials and nanodevices, advances in fundamental theory and modeling, techniques across ranges of scales (from first principles to coarse-graining and phase-field modeling), and education by integrating the cyber toolbox into the fabric of next-generation science and the training of the next-generation modeling community.

Coordinated Activities with Other Agencies and Other Institutions Contributing to This NSI

NIH, CDC, FDA, EPA, and NIST: The National Cancer Informatics Program (NCIP) Nanotechnology Working Group has become involved in this NSI. The objectives of this working group, which are highly relevant to the NKI NSI, are to develop data exchange standards and ontologies, build a community of interest, and eventually enable predictive modeling of nanomaterials activity and rational design of nanomaterials. Working group members include NIH institutes, CDC, FDA, EPA, and NIST, along with representatives from academia, industry, and standards organizations.

NIH and FDA: The NIH Common Fund's Regulatory Science program and FDA have cooperatively funded a project to develop a new method to predict the potential harmful effects of nanoparticles intended for use in clinical applications prior to their testing in humans. Given the increasing emphasis on the use of nanoscale materials for the diagnosis and treatment of disease, if successful, this approach would provide a safety review for potential new therapeutic candidates. [*Also relates to Objective 4.1.*]

NIH, universities, and industry: NCI, NIBIB, and NIEHS work together to manage the Nanomaterial Registry. The goal of the Nanomaterial Registry is to become the definitive cross-disciplinary resource for nanoparticle characterization data for health, toxicity, and industrial concerns. It draws inputs from existing curated databases, including caNanoLab, and currently includes over 1,300 particle entries. Entries are populated on the web portal through curated data extraction using a Minimal Information About Nanomaterials (MIAN) characterization vocabulary architecture. MIANs capture the physicochemical characteristics, biological interactions, and environmental interactions of the given particle. This homogenized vocabulary enables searches and comparisons based on MIAN similarity. [*Also relates to Objectives 4.1 and 4.2.*]

NIOSH, NNI agencies, industry, and universities: In 2015, NIOSH will continue collaborations with the NanoInformatics consortium that includes Federal partners NCI, NIH, NIST, Pacific Northwest National Laboratory (PNNL), and DOD and other public and private sector partners: UCLA, National Nanomanufacturing Network, nanoHUB, RTI International, MIT, and the Nanotechnology Business and Commercialization Association. [*Also relates to Goal 4.*]

NIST, NNI agencies, industry, universities, and international organizations: Multiple NNI agencies are participating in several international forums developing ontologies and nomenclatures for description of ENMs. Federal agency participation in these efforts helps to facilitate data sharing and ensure that agency perspectives and positions are well represented in the resulting documentary standards.

NNI agencies: In collaboration with other agencies, NIST is designing a set of robust tools to assess the quality of the data used to generate the theoretical models as well as the data generated by the computational algorithms implemented by the nanotechnology community. As a result of these activities, the NKI NSI group has drafted a Data Readiness Level (DRL) document[22] highlighting the principles underpinning the data validation metrics for nanotechnology. In addition, agencies participating in the NKI group are designing the framework for a nanotechnology cyberinfrastructure that will lead to the effective capture, curation, and dissemination of data, and computational tools to be used by the nanotechnology community.

Agencies participating in the NSI collaboratively organized a symposium on Nanotechnology Knowledge Infrastructure at Nanotech 2013[23] (May 2013 in Washington DC) and participated in meetings of the NSI group to exchange information and coordinate nanotechnology efforts in this field.

Nanotechnology for Sensors and Sensors for Nanotechnology: Improving and Protecting Health, Safety, and the Environment

Individual Agency Contributions to This NSI

DOD/Air Force: The Materials and Manufacturing Directorate is developing nontraditional sensors to detect biomarkers that indicate fatigue, cognition, and other indicators of human performance. Work is focused on identifying binding agents for these biomarkers as well and the transducer devices that would indicate a binding event and thereby signal the detection of the biomarker. Carbon nanomaterials such as graphene and carbon nanotubes have been demonstrated as potential transducer devices. Efforts are also focused on formulating inks of the binding agents for simple device fabrication.

DOD/Army: Researchers at the Institute for Soldier Nanotechnologies (funded by the Army Research Office and Army Research Laboratory) have explored optoelectronic fibers capable of detecting light, heat, and sound. Unlike traditional approaches for making "smart" fabrics, which center on attaching devices to conventional woven materials, this research focuses on the development of unique fibers that are themselves optoelectronic devices. The fibers are already used for over 2,400 endoscopic medical procedures a month across about 1,000 hospitals. Future applications could include combat identification, infrared communications, sniper detection, medical imaging, and blast and blunt impact monitoring.

DOD/DARPA: Natural biological sensors often display high sensitivity, selectivity, and low false alarm rates while being fabricated and operated in dirty, noisy natural environments. Examples of such sensors include the highly efficient energy transfer properties of photosynthesis in plants, bacteria, and algae; magnetic field sensing used by some birds for navigation; and the ability of some animals to detect odors at the single-molecule level. Attempts to emulate these sensors synthetically through structure or chemistry have not fully met expectations. Recent evidence suggests that some biological sensors exploit nontrivial quantum mechanical effects to produce macroscopic output signals. The DARPA physics in biology program is laying the foundation for novel sensor designs by challenging the long-held research community view that biological sensors utilize primarily classical physics. Examples of recent accomplishments under this multidisciplinary program include the first evidence for coherent quantum transport in photosynthetic systems and initial experimental observations of quantum effects in insect olfaction.

NASA: NASA supports work in the development of sensors for the detection of chemical and biological species for use in planetary exploration, vehicle performance monitoring, and astronaut health management. In 2013, NASA researchers utilized porous tin oxide nanorods fabricated from multiwall carbon nanotubes in sensors for the detection of methane over a temperature range of 20 °C to 500 °C. A carbon nanotube trace gas sensor developed by NASA Ames researchers was awarded the 2012 NASA Government Invention of the Year. Work on the development of autonomous chemical-biological sensor platforms based on carbon nanotube and nanofiber sensing elements will be initiated in 2014 with funding from the Space Technology Mission Directorate's Game Changing Development Program for potential use in closed-loop life support systems and planetary rovers.

NIH: NIH has been directly and intensively engaged in this NSI from its inception. NIH supports a significant amount of research in biosensor and nanopore sequencing device development, and in biomarker discovery and validation (NSI Thrust 1).[24] NIH further supports research on nanomaterials toxicity and lifecycle issues through investigator-initiated grants as well as NCI's Nanotechnology Characterization Laboratory, which provides protocols for physicochemical characterization of nanomaterials and assay techniques for elucidation of nanomaterial-host interactions (NSI Thrust 2). NCI is sponsoring a small (~$150,000) contract solicitation to support development of techniques and materials to facilitate early-stage development of nanotechnology-enabled sensors as a direct result of its involvement in this NSI.

NIOSH: In support of the sensors signature initiative, NIOSH has invested in the development and/or testing and evaluation of direct-reading instruments capable of detecting and measuring airborne nanoparticles. NIOSH will continue to support the development of guidelines and voluntary consensus standards for identification of sensor needs specific to the objective of realistic applications in complex workplace situations. [*Funding for this activity is reported by NIOSH under PCA 5; also relevant to Objective 4.1.*]

NIST: NIST is developing a variety of measurement methods to characterize nanosensors and nanomaterials, and it is creating nanosensors and nanomaterials for measurement applications and standards, including low-power, portable semiconductor nanowire sensors for environmental and biomedical applications and advanced chemical and biochemical sensors enabled by nanostructured materials. Reference materials developed at NIST support the validation of new sensing platforms. NIST's nanotechnology user facility, the CNST, provides access to the tools and processes needed to accelerate the commercialization of nanosensor systems, along with novel nanofabricated fluidic devices and advanced microscopy systems created and customized to control and measure nanoparticles and biomolecules. Research at NIST addressing information quality, integrity, and usability will contribute to the reliability and security of sensor networks and data analysis. NIST's complementary nanotechnologyrelated environmental, health, and safety (nanoEHS) programs support the safe manufacture, use, and disposal of ENMs and nanotechnology-enabled products.

NSF: Through its Nano-Biosensing, Biophotonics, and other programs, NSF supports development of novel sensitive, discriminative, low-cost, and easy-to-operate biosensing systems; innovative ideas in the development of novel biorecognition strategies; multifunctional nanomaterials and interfaces with predefined physical, chemical, or biological characteristics for biosensing applications; and fundamental study of biomacromolecule confinement and orientation at the micro- and nanoscale interfaces for biosensing applications. NSF also supports the development of sensors to detect engineered nanoparticles in a variety of environmental matrices.

USDA/NIFA: The agency utilizes multiple funding authorities to support development of nanobiosensors for more sensitive, specific, and robust detection of pathogens, toxins, and contaminants in food to ensure food safety and biosecurity. NIFA also continues to support the development of sensors for monitoring the environmental stresses in crop and livestock production, and the distribution of agricultural chemicals in fields to provide critical information for precision agriculture. In the forefront of sensors for nanotechnology, NIFA grantees participated and contributed to the International Life Sciences Institute NanoRelease project on detection, identification, and characterization of engineered nanomaterials in foods and packages. NIFA has been supporting and providing leadership to a multistate research committee (NC1194: Nanotechnology and Biosensors[25]). The committee has been effective in advancing nanoscale science and engineering for nanosensor development and commercialization.

Coordinated Activities with Other Agencies and Other Institutions Contributing to This NSI

DOD/Air Force, industry, and universities: The Materials and Manufacturing Directorate activities in nanotechnology for sensors are highly coordinated with AFRL Human Effectiveness and Sensors Directorates as well as the AFRL-sponsored Nano-Bio Manufacturing Consortium. Numerous universities including Northeastern University, University of Cincinnati, and others contribute to the work in addition to several companies, including MC10, General Electric, and Brewer Science.

NASA and DHS: A carbon nanotube-based trace gas sensor developed by NASA researchers has been integrated into a smart phone platform with support from DHS. Successful field testing of this sensor as a portable system for the detection of toxic gases released during a fire and in a network of sensors to establish safer perimeters was performed with the Los Angeles Fire Department in late 2011. These sensors are currently being developed for detection of chemical contaminants in water for closed-loop life support systems.

NIH, FDA, and NIST: NIH is interacting with other participating agencies to advance the goals of this NSI. NCI staff members are engaged in discussions with staff in the Center for Devices and Radiological Health within FDA about issues in identifying appropriate standards for use with nanotechnology-enabled sensors and devices and how best to evaluate the performance of these devices. NIST participants in the NSI have also weighed in on these discussions, and this topic will be included in a workshop planned for 2014.

NIOSH, NIH, FDA, NIST, DOD, NASA, NSF, and EPA: NIOSH is collaborating with other agencies on an initiative to develop nanotechnology-enabled sensors and sensors for nanomaterials for EHS research applications. The objective of the initiative will be to develop and commercialize sensors that will enable more specific tracking of engineered nanomaterials throughout their life cycles, thereby supporting the need for private-sector companies to demonstrate responsible development. NIOSH plans specific applications in the areas of detection of nanomaterials in biologic systems to evaluate and predict biological behavior and translocation between organ systems; more sensitive and specific detection of worker exposures to nanomaterials with an ultimate outcome of real-time detection; and sensing of biomarkers of nanomaterial exposure and early response to support worker surveillance. [*Also relates to Objective 4.1.*]

NNI agencies: Agencies participating in the NSI collaboratively organized a symposium on Nanosensing at Nanotech 2013[26] (May 2013 in Washington DC) and participated in meetings of the NSI group to exchange information and coordinate nanotechnology efforts in this field.

Agencies participating in this NSI reviewed responses to a Request for Information posted in the Federal Register[27] to solicit community input regarding specific needs for the accelerated development and commercialization of nanosensors. Input received with respect to standards, testing, manufacturing, commercialization, and regulation will inform the design of a nanosensors workshop planned for 2014.

Goal 2: Foster the Transfer of New Technologies into Products for Commercial and Public Benefit

The purpose of Goal 2 is to establish processes and resources to facilitate the responsible transfer of nanotechnology research into practical applications and capture its benefits for national security, quality of life, economic development, and job creation. Several factors are necessary to successfully commercialize any new technology. Scalable, repeatable, cost-effective, and high-precision manufacturing methods are required to move the technology from the laboratory into commercial products. Investments by both the public and private sectors are needed to shepherd technologies to maturity. Maximizing the benefits of nanotechnology developments to the U.S. economy requires efforts to remove barriers to global commercialization and an understanding of the potential markets for a given product. Goal 2 encompasses four objectives that detail how the NNI will focus its resources and broaden its engagement with academia, industry, and the international community to reach this goal.

The NNI member agencies also have a number of activities uniquely targeting technology transfer and commercialization, for example, workshops to obtain input from industry and academia, SBIR and STTR programs to fund innovations in small businesses, and cutting-edge research infrastructure for use by all nanotechnology researchers, including those from industry.

Objective 2.1 – Assist the Nanotechnology-Based Business Community in Understanding the Federal Government's R&D Funding and Regulatory Environment

Individual Agency Contributions to Objective 2.1

DOC/BIS: The Bureau of Industry and Security (BIS) advances U.S. national security, foreign policy, and economic objectives by ensuring an effective export control and treaty compliance system, and by promoting continued U.S. leadership in strategic technologies, including nanotechnology. BIS accomplishes its mission by maintaining and strengthening adaptable, efficient, and effective export control and treaty compliance systems. BIS export control outreach and education constitute the first line in the bureau's contact with U.S. exporters and provide guidance and transparency to new and experienced exporters regarding the Export Administration Regulations (EAR). BIS's activities include seminars, webinars, teleconferences, and on-location panel sessions at various conferences. Additionally, one-on-one counseling assistance is provided on both the East and West Coasts for extended hours of daily operation. Over the past few years, BIS has developed capabilities to offer training online and via interactive webinars. Through these programs, BIS provides guidance on regulations, policies, and practices and helps to increase compliance with U.S. export control regulations. BIS has also developed an introductory series of easy-to-use training modules. This service offers exporters and re-exporters a cost-saving mechanism to learn about U.S. dual-use export controls. BIS services are particularly useful for small and medium-sized businesses that operate with limited compliance resources.

FDA: FDA routinely presents its scientific considerations and regulatory approach to nanotechnology to its stakeholders, including the nanotechnology-based business community, through participation in various public forums. For example, in 2014 FDA is participating in the Product Quality Research Institute Workshop on Nanomaterial Drug Products, which includes two sessions dedicated to regulatory issues related to nanotechnology applications in pharmaceutical products. FDA also encourages and provides for industry consultations such that the potential effects of use of nanomaterials on the safety, effectiveness, or regulatory status of a product can be identified and addressed on a case-by-case basis. These activities directly assist nanotechnology-based businesses in understanding the FDA regulatory environment while also ensuring the safety and, as applicable, effectiveness of nanotechnology products.

IC/DNI: The National Reconnaissance Office has made significant progress toward Objective 2.1 [*progress also relevant to Objective 2.2*], including the following examples: the first compression-molded carbon nanotube composite parts; an optical mirror using carbon nanotube material to prevent print-through; lithium ion carbon nanotube electrodes capable of 300 watt-hours per kilogram (a two-fold increase in energy capacity); transfer to production of carbon-nanotube-based heating elements; environmental testing of carbon nanotube materials by the U.S. Navy for maritime use; transfer of carbon nanotube sheets for enhanced ballistic protection for soldiers and high-value assets; transfer of carbon nanotube sheets for lightweight flooring for commercial aircraft.

NASA: In 2013, the NASA Office of Chief Technologist developed its TechPort website to manage NASA's advanced technology investments and make information on them more accessible. The website contains information on NASA's efforts in nanotechnology R&D, including work funded at NASA Centers as well as contracts with industry, and the products of these activities. TechPort is currently being beta-tested internally at NASA prior to public release.

NSF: The agency supports programs to promote university-industry interaction (Grant Opportunities for Academic Liaison with Industry or GOALI, Industry/University Cooperative Research Centers or I/UCRCs, and Partnerships for Innovation or PFI), translational research (Engineering Research Center or ERC and the Innovation Corps or I-Corps), and small business innovation (SBIR/STTR) in nanotechnology.

USDA/NIFA: The USDA SBIR program has continued to support nanotechnology R&D aiming at commercialization. The awards have included nanotechnology-enabled sensor technologies for detection of microbial pathogens and insects and crop environmental stresses, nanoscale encapsulation to deliver anti-pathogenic agents, and nanocellulose composites to improve polymer product functional properties.

USPTO: The transfer of new nanotechnology-related technologies into products for commercial and public benefit depends on effective mechanisms that protect new ideas and investments in innovation and creativity. USPTO is at the cutting edge of the Nation's innovation system, providing intellectual property policy advice and guidance to the Executive Branch. The agency has put in place several initiatives to keep pace with the rapid advances being made in nanotechnology. The agency continues to provide in-depth nanotechnology-specific training events for patent examiners as well as to foster communication among examiners across multiple disciplines. The USPTO has a subset of patent examiners across all technology disciplines who serve as points-of-contact to assist other examiners with nanotechnology issues related to patent examining.

Coordinated Activities with Other Agencies and Institutions Contributing to Objective 2.1

NSF and NNI agencies: The agency will continue to co-sponsor nanotechnology regional alliances conferences together with other agencies by providing additional funding for academic participants and setting a long-term perspective. Among these are the annual National Nanomanufacturing Network conference and the NNI-sponsored series of workshops on regional, state, and local nanotechnology initiatives.

Objective 2.2 – Increase Focus on Nanotechnology-Based Commercialization and Related Support for Public-Private Partnerships

Individual Agency Contributions to Objective 2.2

DOD/Air Force: The Materials and Manufacturing Directorate supports nanotechnology-based commercialization through partnerships/collaborations with small businesses as well as through the public-private Nano-Bio Manufacturing Consortium, which is sponsored by several laboratory directorates. This consortium aims to develop shared platforms for commercializing nanotechnology and biotechnology related to human performance assessment. This three-year effort is based on government/industry /university cost-sharing, with technical efforts selected by a team representing member institutions. The initial platform being developed is a wearable human performance assessment patch.

DOD/Army: Commercialization of nanotechnologies must consider the potential liabilities associated with releases from the value chain and throughout the product life cycle. Starting in 2014, the Army is collaborating with military industry partners to develop environmental life cycle analysis approaches for nanotechnologies. These collaborations are used to identify areas of potential nanoparticle release during production and future product liabilities during use and end of the technology life. Cooperative partnerships between the Army and industry will result in more rapid transition of technology to production and environmentally sustainable practices. [*Also applies to Objective 4.4.*]

DOD/DARPA: DARPA STARnet is already leading the way with a unique public-private partnership. DARPA and multiple member companies are working in a true partnership, sharing the costs and management to fund breakthrough university research to better understand nanoscale physics and materials science, and to realize devices and circuits with unprecedented performance. The private companies and universities have all agreed to the same intellectual property (IP) rights (DARPA has usual and customary IP rights for DOD). The private companies and DOD lab experts are encouraged to collaborate and to work with university researchers under the program, sharing the research with all, and gaining knowledge and expertise along the way.

DOD/Navy: Under two separate Navy programs to advance development and commercialization of nanomaterials with unprecedented properties, universities and industry are conducting cooperative research and development. In the first, Raytheon has been teamed with several universities to develop a new generation of infrared (IR) transparent ceramics for extreme environments such as those experienced by missile IR domes. This work has resulted in development of thermal shock-resistant domes with unprecedented optical properties, which have been accepted for transition to the AIM-9X Sidewinder missile. A pilot plant is being constructed, and further development is being funded by ONR, DOD Manufacturing Technology, and the Air Force. Under the second program, Integran, Inc., has been paired with Rutgers University to develop hybrid structures consisting of graphene-reinforced polymers coated with electrodeposited nanostructured alloys. The structures exhibit remarkable strength, toughness, and ballistic properties and are being tested by Pratt & Whitney for aircraft engine applications.

DOD/OSD (Office of the Secretary of Defense): Since 2011 DOD and other U.S. Government agencies have partnered with industry in establishing the world's first commercial-scale CNT yarn, sheet, and tape production capability. Through Defense Production Act, Title III authorities, pilot production will be established in 2014. In 2015, capacity expansion will continue in order to meet low-rate initial production requirements. In 2013 the development of "super fuels" increased production by an order of magnitude, reducing costs. Commercially, CNT spacecraft data cables, utilizing yarns and tapes, have achieved technology readiness level (TRL) 6. Several national asset platforms have been identified for CNT product insertion in 2014–2015. CNT sheet-based armor products have achieved TRL 6, realizing weight savings of 10–19% in hard armor and 30% in soft armor, while maintaining or even increasing protection and mobility. Potential benefits extend from ground troops to improving survivability and reducing operating costs of legacy aircraft, helicopter, sea, and ground transportation systems.

DOE: In order to transfer new technologies into the market, EERE-SETP supports the SunShot Incubator Program. The Incubator Program provides early-stage assistance to help startup companies cross technological barriers to commercialization while encouraging private sector investment. Projects that relate to nanotechnology include concepts such as silicon nanowires to increase power density and reduce costs. Additionally, SETP funds three manufacturing development centers through the Photovoltaic Manufacturing Initiative and the National Renewable Energy Laboratory's Process Development Integration Laboratory.

The EERE solid-state lighting program supports research and development of promising SSL technologies through annual competitive solicitations. Manufacturing R&D projects accelerate SSL technology adoption through manufacturing improvements that reduce costs and enhance quality of LEDs and OLEDs. Research includes the development of high-speed, high-resolution, nondestructive test equipment with standardized test procedures and appropriate metrics within each stage of the value chain for semiconductor wafers, epitaxial layers, LED dies, packaged LEDs, modules, luminaires, and optical components. Also included is the development of manufacturing equipment enabling high speed, low cost, and uniform deposition in state-of-the-art OLED structures and layers.

NASA: NASA continues to support the development and commercialization of nanotechnology-related products through the SBIR and STTR programs. In 2012 NASA invested over $20 million in nanotechnologyrelated SBIR and STTR projects, including funding of R&D related to the development of advanced nanoscale materials, electronics, sensors, propulsion systems, and propellants.

NIH: In 2012, NCI initiated the TONIC consortium to accelerate the translation and development of nanotechnology solutions for the early detection, diagnosis, and treatment of cancer. This partnership model has several goals, including providing NCI-supported researchers insight into industry needs in technology platforms and drug targets, promoting collaborations between academic investigators and industry partners on precompetitive and late-stage programs, providing TONIC members the opportunity to interact with regulatory authorities and NCI's Nanotechnology Characterization Laboratory, and serving as a sustained forum for exchange of ideas on nanotechnology. TONIC's membership continued to grow in 2013 with the addition of several small companies.

TONIC activities include sponsoring symposia at national meetings and hosting stand-alone workshops, including one devoted to the enhanced permeability and retention (EPR) effect, an important feature of nanomedicine delivery that strongly influences patient response. This workshop has resulted in the formation of a nanodrug working group with members from academia and large pharmaceutical companies to work on clinical proposals/work ideas for effective imaging approaches to study EPR activity in patients. The working group has also recognized and is addressing the fundamental limitations and gaps in preclinical tumor models in recapitulating characteristics of solid tumors in patients.

The NCI Alliance for Nanotechnology in Cancer remains focused on promoting clinical translation and commercialization of research results. Eight Alliance-affiliated therapeutics are being tested in 17 cancer-related clinical trials, and five devices and instruments have started clinical trials or Institutional Review Board (IRB) approved studies in humans. Alliance researchers are partnering with more than 70 companies to translate their discoveries to the clinic or marketplace. Commercially available products include:

- The GENISYS4 small animal PET (positron emission tomography) scanner for research use, a novel PET probe, and the ELIXYS radiosynthesizer for laboratory production of PET probes.

- MagVigenTM, MaxVigenTM, AuVigenTM, and MyQuVigenTM multifunctional and biocompatible nanoparticles for sample separation, molecular imaging, *in vitro* diagnostic or therapeutic applications.
- More than 360 distinct gold nanoparticle-based SmartFlareTM detection probes for research use.

NIST: NIST senior management, in coordination with NIST's Technology Partnership Office, is working to promote commercialization, technology transfer, and entrepreneurship in nanotechnology broadly across NIST's laboratory programs through a revised patent policy, training, additional in-house patent and technology transfer resources, and leadership of the Federal Laboratory Consortium for Technology Transfer.

NSF: NSF sponsors public-private partnerships (e.g., the Nanoelectronics Research Initiative and I-Corps) and centers that advance collaboration with industry (e.g., the Nanosystems ERCs and NSECs). NSF sponsored the Nanotechnology Undergraduate Education project "NUE: NanoTRA–Texas Regional Alliance to foster 'Nanotechnology Environment, Health, and Safety Awareness' in tomorrow's Engineering and Technology Leaders." NSF also has supported a survey on nanomanufacturing through sponsorship of a study entitled "Nanotechnology and Commercialization—Achieving Sustainable Nanoproducts." This study, conducted by the National Center for Manufacturing Sciences, is a follow-up to previous similar studies sponsored by NSF. It is a 30-question online survey and includes interviews.[28]

Coordinated Activities with Other Agencies and Institutions Contributing to Objective 2.2

DOD/Air Force and industry: The Nano-Bio Manufacturing Consortium is a collaboration between 10 industrial and academic members, in addition to the AFRL Materials and Manufacturing, Sensors, and Human Effectiveness Directorates, and the FlexTech Alliance, which manages efforts on behalf of AFRL.

DOD/Army and universities: In a joint venture, the Institute for Soldier Nanotechnologies and the Army Research Laboratory collaborate on research in novel applications of chemical vapor deposition (CVD). This technology is enabling the deposition of thin (nanometer to micrometer) conformal coatings of heat- and chemically sensitive polymers on diverse substrates. Additionally, ISN collaboration with the Natick Soldier Research, Development and Engineering Center has resulted in the use of gentle CVD to strongly increase the electrical conductivity of functionalized coated surfaces for the detection of pathogens in food.

DOE and NSF: NSF (Directorate for Engineering) and DOE (EERE-SETP) co-sponsor the Quantum Energy and Sustainable Solar Technologies (QESST) Engineering Research Center, which works with companies to advance technologies toward commercialization.

IC/DNI, DOD, and industry: Nanomanufacturing of bulk carbon nanotube sheets, yarns, and tapes continues to attract commercial industry attention, specifically from the aerospace industry (Boeing, Airbus, Bombardier, and Embraer).

The Office of the Secretary of Defense's Defense Production Act Title III has added an additional $14.6 million appropriated in 2013. These funds will assist in building out the world's first bulk carbon nanotube pilot plant in Merrimack, NH. A manufacturing process for compression-molded carbon nanotube composite parts has been developed and demonstrated. Scale-up to commercial scale is scheduled to begin in 2014.

NIH, FDA, NIST, and industry: The Nanotechnology Characterization Laboratory is a formal interagency collaboration of NCI's Alliance for Nanotechnology in Cancer with FDA and NIST. NCL's efforts have helped to drive private-sector investment in nanomedicines. This past year, major pharmaceutical firms, such as AstraZeneca, invested in nanomaterials evaluated by NCL. Companies with NCL-tested nanomedicines now have over $1 billion in licensing agreements with the pharmaceutical industry.

The NCL has standardized more than 40 *in vitro* assays for nanomaterial characterization. These assays have been validated for a variety of nanomaterial types and undergo continual revision to ensure they meet FDA regulatory requirements. New assays are added every year and are made freely available to the public via the NCL website (ncl.cancer.gov/working assay-cascade.asp). In collaboration with NCL, NIST led the development of 11 physicochemical characterization protocols listed on this website. Many of these assays have also been compiled into a recent methods book, *Characterization of Nanoparticles Intended for Drug Delivery.*[29] The NCL's three-tiered assay cascade includes physicochemical, *in vitro*, and *in vivo* characterization. The NCL's physicochemical characterization of nanomaterials goes well beyond basic measurements of size and surface charge. NCL's routine characterization also includes batch-to-batch consistency evaluation, measurement of drug loading, confirmation of targeting ligand conjugation, quantitation of surface ligands, and nanoformulation stability assessment. *In vitro* analysis includes sterility and endotoxin quantification, something many researchers often overlook, and analysis of hematological compatibility and immune cell functions using human whole blood. NCL's *in vivo* capabilities include toxicology, immunotoxicology, drug metabolism, pharmacokinetics, efficacy, and imaging studies.

NIST, NSF, industry, and universities: NSF and NIST continue their support of the Nanoelectronics Research Initiative, which announced a second major phase and annual funding for three multi-university research centers in 2013. NRI is a public-private partnership supporting the development of future computing devices and includes collaborative research among NIST and NRI researchers from industry and universities.

USDA/FS and industry: In 2013, the Forest Service formed a public-private partnership with the U.S. Endowment for Forestry and Communities to advance cellulose nanomaterials research. This new partnership, announced in December 2013, will provide funding for collaborative CN R&D and develop new stakeholders in support of commercializing cellulose nanomaterials. The Forest Service, in collaboration with NNCO, plans to hold a workshop to identify gaps in commercializing CN in 2014. FS and NNCO organizers plan to invite CN users to the workshop. Forest Service will use the outcomes of the workshop to inform nanotechnology R&D planning.

Objective 2.3 – Promote Broader Accessibility and Utilization of User Facilities, Cooperative Research Centers, and Regional Initiatives to Accelerate the Transfer of Nanoscale Science from Lab to Market

Individual Agency Contributions to Objective 2.3

DOE: The Office of Science continues to operate five Nanoscale Science Research Centers (NSRCs), national user facilities for interdisciplinary research at the nanoscale, serving as the basis for a national program that encompasses new science, new tools, and new computing capabilities. Each center has particular expertise and capabilities in selected theme areas, such as synthesis and characterization of nanomaterials; catalysis; theory, modeling and simulation; electronic materials; nanoscale photonics; soft and biological materials; imaging and spectroscopy; and nanoscale integration.

EERE-SETP supports solar energy activities at the National Renewable Energy Laboratory though an annual operating plan that contains a collection of efforts related to nanotechnology. One example is research utilizing nanomaterials to enhance transparent conducting materials and efforts to increase the resolution of measurement techniques to electrically and optically probe grain boundaries, interfaces, and structures. The BRIDGE program supports the development of advanced characterization and process development tools at a number of DOE-SC-funded user facilities.

NIH: The NCI Alliance for Nanotechnology, the NHLBI Programs of Excellence in Nanotechnology, and the NIH Roadmap for Nanomedicine Centers all follow a cooperative research model. [*See Objectives 1.1 and 2.2.*]

NIST: CNST has released a major update to its NanoFab website and deployed a new laboratory equipment management system developed from the ground up to enhance the accessibility and utilization of the facility. The updated website includes significant new content describing how to apply to and work in the NanoFab, status information and news, an FAQ, and a complete redesign of the site's navigation. The concurrently launched web-based laboratory equipment management program allows onsite NanoFab users to make tool reservations, controls access to tools in the NanoFab, and streamlines logistics and communication. The CNST is also working to develop capabilities for remote access to the NanoFab in order to promote the use of nanofabrication and measurement by researchers who do not have ready access to such facilities.

NSF: NSF supports and has re-competed the Network for Computational Nanotechnology in 2013 and the National Nanotechnology Infrastructure Network in 2014. Both are user facilities with nationwide scope and availability.

Coordinated Activities with Other Agencies and Institutions Contributing to Objective 2.3

NSF, DOD, NIST, and industry: NSF supports the National Nanomanufacturing Network (NNN), which includes the NSF NSECs and non-NSF centers in collaboration with the Department of Defense, the National Institute of Standards and Technology, and industry partners in an alliance to advance U.S. strength in nanomanufacturing. NSF plans to create a database of all federally supported user facilities.

Objective 2.4 – Actively Engage in International Activities Integral to the Development and Responsible Commercialization of Nanotechnology-Enabled Products and Processes

Individual Agency Contributions to Objective 2.4

DOC/BIS: BIS has developed and published two comprehensive documents on its Export Management and Compliance Programs (EMCPs), which are programs that can be established to manage export-related decisions and transactions to ensure compliance with the Export Administration Regulations and license conditions. Specifically, BIS has published

"Compliance Guidelines: How to Develop an Effective Export Management and Compliance Program and Manual" and "EMCP Audit Module: Self-Assessment Tool." Additionally, BIS conducts regularly scheduled outreach seminars dedicated to the topic "How to Develop an Export Management and Compliance Program" and has added the webinar "Elements of an Effective Export Compliance Program" to its menu of online training offerings.

DOD/Navy: In June of 2012, the Chief of Naval Research (CNR) made the first-ever trip by a CNR to Israel. During the trip, in a meeting with the Israeli Ministry of Defense and Directorate of Defense Research and Development—the Israeli defense R&D organization also referred to as MAFAT—it was agreed that annual executive level meetings would be held and two initial working groups would be created, on advanced materials and on robotics and autonomous systems. On 20 May 2013, a working group meeting on advanced materials was held in Arlington, VA, co-chaired by representatives of ONR and MAFAT. The participants identified several topics in nanotechnology and advanced nanomaterials for potential cooperative development. Further partnership meetings have been held since June 2012, and the CNR is expected to make a return trip to Israel in July 2014. A follow-on workshop with a nanomaterials focus is being planned for March 2014.

DOD/Air Force: To seed innovation in areas that are relevant to Air Force capabilities and that can potentially benefit from research and technology development in nanomaterials and devices, Air Force researchers have engaged the international community to encourage work in specific areas and characterization against metrics that indicate technology potential. In areas such as graphene nanodevices, nanomagnetic materials, and optical nanoantennas, the Materials Directorate has established recognized metrics that can facilitate wide technical evaluation on the part of researchers and both guide and identify research with technological potential in these areas. This process can also identify classes of materials that may provide unique advantages over a technology area, such as nanopatterned self-assembled metallic structures, and allow for focused nanomanufacturing capability development to support broad-based commercialization.

DOS: The agency actively engages in international activities integral to the development and responsible commercialization of nanotechnology-enabled products and processes.

Coordinated Activities with Other Agencies and Institutions Contributing to Objective 2.4

DOD/Army, EPA, NIST, and international organizations: The Army Corps of Engineers, in cooperation with EPA and NIST, is exploring methods and approaches to assess the environmental safety of nanotechnologies that are needed to support scientifically based management and regulatory decisions. Starting in 2014, in response to the 2013 National Academies review and the NNI 2013 stakeholder workshop on the potential risks of nanotechnology, the Army is collaborating with EPA, NIST, Germany, Austria, Canada, the United Kingdom, and industry to develop standardized testing decision trees and screening methodologies to assess release, fate and transport, and toxicity of nanoparticles. Methods for assessing release of nanoparticles are being developed through the International Life Sciences Institute and the Organization for Economic Co-operation and Development (OECD). [*Also applies to Goal 4.*]

DOD and NNI agencies: DOD international activities are coordinated within the department through the military department and DOD agency international programs offices and the Pentagon. When appropriate, international efforts are coordinated with other Federal agencies, primarily through NSET and its global issues interest group.

DOS, NNI agencies, and OECD: OECD's Working Party on Nanotechnology (WPN) provides advice on emerging policy issues of science, technology, and innovation related to the responsible development of nanotechnology. The Department of State serves as Vice-Chair on the WPN bureau and coordinates NNI agency participation in ongoing projects, including "Key Statistics and Indicators for Nanotechnology," "Commercialization of Nanotechnology Research," and "Skills and Education for Nanotechnology." For its 2013–2014 program of work and budget, the WPN is considering policies to understand and monitor the societal and economic impacts of nanotechnology; linking nanotechnology to broader policies for sustainable and green growth; and policy issues raised by the convergence of nanotechnology with other technologies.

NIOSH and international standards organizations: In 2015, NIOSH will continue to play an active role in the International Organization for Standardization (ISO) technical committee (TC) 229 – Nanotechnology and ASTM International technical committee E56 – Nanotechnology, both of which focus on development of consensus standards for industry and commerce. A critical focus in 2015 will be the execution of an international interlaboratory study to develop methods for determination of nanomaterial size for high production volume nanomaterials including titania and carbon black. Results will be disseminated as an ISO Technical Report. Additionally, through ASTM International, NIOSH is working to develop a standard guide on sample preparation for particle sizing and a standard guide on detection and characterization of silver nanomaterials in commercial products.

NIOSH, NIST, and international organizations: NIOSH is collaborating with NIST and the National Research Council of Canada (NRC-Canada) to develop and qualify nanoscale reference materials to support measurement quality. In 2014, NIOSH collaborated with NIST on surface area measurements for NIST Standard Reference Material 1898: Titanium Dioxide Nanomaterial, which is now commercially available. In 2015, collaborations with NRC-Canada will continue on qualification of two cellulose nanocrystal reference materials and one single-wall carbon nanotube reference material. NIOSH and other agencies will use these reference materials to improve measurement quality, which enables new areas of research and strongly supports the critical and necessary intersection of materials measurement scientists with toxicology and occupational health professionals.

NIOSH and other NNI agencies: NIOSH, in collaboration with other NNI agencies, contributes in the areas of human health, exposure assessment, risk assessment, and risk management and controls by participating in multiple communities of research (CoR) as part of the U.S.-European Union (EU) effort to bridge EHS research. In addition to active participation in the CoRs, NIOSH has actively engaged research institutes in Canada, Brazil, Japan, China, Sweden, the United Kingdom, and Russia in several areas of nanotechnology EHS research collaboration.

NIST and standards developing organizations: NIST staff members are significant contributors to the development of international consensus standards that are integral to the development and responsible commercialization of nanotechnology-enabled products and processes. NIST provides measurement and characterization expertise and leadership in the development of standards in organizations such as ASTM International, the International Electrotechnical Commission (IEC), and the ISO, and also supports U.S. Government agency participation in organizations such as the OECD Working Party on Manufactured Nanomaterials (WPMN). Through engagement in organizations that develop international standards, NIST has been instrumental in the development of nanotechnology standards in the form of terminology specifications and standards, technical reports and guidance documents for measurement methods, and validated test methods to quantitatively determine physicochemical and toxicological properties of ENMs with consistent, verifiable results.

NIST continues to raise awareness of standards development activities and their impact through engagement with other agencies and in interagency groups such as the NEHI Working Group and the nanotechnology subgroup of the White House Emerging Technologies Interagency Policy Committee, in collaboration with the NNI's global issues and standards coordinators.

NSF and international organizations: NSF continues to participate in OECD, ISO, and other international forum activities.

USDA/FS and international standards organizations: To remove barriers for commercialization, the Forest Service continues to lead international standards development in cellulose nanomaterials. In 2013, CN standards development reached several milestones. In 2014, Forest Service and TAPPI plan to submit the CN terminology standard to ANSI to become an ANSI standard and prepare an ISO standards-development proposal based on the TAPPI standard. In ISO TC 229, as a result of an effort led by Forest Service experts working in collaboration with international collaborators, ISO standards in terminology, characterization, and material specification for CN were included in the resolution of the 2013 TC meeting.

The CN Task Group in ISO TC 6 (Pulp, Paper and Board) has continued to explore characterization standards needs in 2013 and 2014. Working in coordination with the United States, the Canadian Standards Association is in the final steps of preparing a draft for a CN characterization standard.

USPTO and other nations' patent offices: USPTO is moving to a new patent classification system, the Cooperative Patent Classification (CPC), which is jointly managed by USPTO and the European Patent Office (EPO). This new classification system is based on an internationally used patent classification system (IPC), which contains most of the world's patent documents.

As USPTO moves to this new classification system, there will be a more harmonized, internationally consistent classification of nanotechnologyrelated patent documents.

Goal 3: Develop and Sustain Educational Resources, a Skilled Workforce, and a Dynamic Infrastructure and Toolset to Advance Nanotechnology

Fundamental to the successful development of nanotechnology is the continued development of the infrastructure necessary to support this effort. A substantial investment, strengthened by interagency cooperation and collaboration through the NNI, is needed to develop the talent and facilities necessary to achieve the other NNI goals of advancing a world-class R&D program (Goal 1), fostering the transfer of new technologies into products for commercial and public benefit (Goal 2), and supporting responsible development of nanotechnology (Goal 4).

This goal encompasses three objectives[30] that detail how the NNI will responsibly engage and educate the public and the workforce regarding the opportunities that nanotechnology offers and the skills it requires, along with providing the needed access to advanced facilities and tools. Education is among the chief objectives of NNI-funded university research. In addition, specific programs target K–16 education, improve nanotechnology curricula in U.S. schools and universities, and educate the public about nanotechnology. The NNI continues to sustain, maintain, and upgrade its extensive network of research centers, user facilities, and other infrastructure for nanotechnology research, a key element of the original NNI strategy.[31]

Objective 3.1 – Sustain Outreach and Informal Education Programs in Order to Inform the Public About the Opportunities and Impacts of Nanotechnology

Individual Agency Contributions to Objective 3.1

DOD/DARPA: There are numerous formal and informal efforts within the DARPA STARnet program and its six multi-university centers for education and outreach. The program establishes a vibrant academic environment with frequent technical exchanges through seminars and webcasts that are open to the public. Research results are frequently published in technical and nontechnical journals, magazines, and books. Center directors and researchers have outreach to companies, government labs, and present their work in public meetings.

IC/DNI: The National Reconnaissance Office has competitively selected a 2013 Intelligence Community (IC) postdoctoral researcher to work on nanotechnology challenges in carbon nanotube chirality control and separation by chirality. The 2014 IC postdoctoral proposals include three nanotechnology topics: (1) thermoelectric devices based on carbon nanotubes to harvest heat from data centers and spacecraft electronics, (2) use of gallium antimonide compounds to produce quantum dot solar cells with 41 to 47% efficiencies, and (3) boron nitride nanotube development. An academic institution has been selected competitively to research boron nitride nanotube materials in 2014 under an innovation seedling contract.

NSF: Two Centers for Nanotechnology in Society with significant educational components are funded at Arizona State University (ASU) and the University of California, Santa Barbara (UCSB). The CNS-UCSB supports interdisciplinary education across its research activities in various ways. Graduate students and postdoctoral researchers associated with one or more of the center's research thrusts are being mentored by their CNS advisors. Previous graduate students and postdoctoral researchers received jobs in academia and in the private sector in areas related to societal implications of nanotechnology. CNS outreach activities include engaging with the Centers for Environmental Implications of Nanotechnology at UCSB and with policy groups and nongovernmental organizations (NGOs); these activities promote increased awareness of societal issues within the nanoscale science and engineering community. Public engagement efforts include media interviews, speaker series, NanoDays,[32] web-based materials, and public presentations. CNS-ASU has been offering an immersive training session for nanoscale science and engineering graduate students in Washington DC called Science Outside the Lab. The CNS has established an annual, two-week "Winter School" at ASU for doctoral students and postdocs. Looking ahead, the CNS plans to focus on recruiting, funding, and other resources for the Winter School on students from underrepresented groups. It also plans to develop and implement targeted recruiting efforts for a new Graduate Certificate in Responsible Innovation and a minor in Science & Technology Policy, and it will organize and host recruiting events with the Hearing Research Center, the ASU American Indian Studies Program, and the American Indian Policy Institute, among others.

The NSF-supported Nanotechnology Applications and Career Knowledge (NACK) and the Nanoscale Informal Science Education Network (NISE Net) are two national networks serving the nanotechnology technological community and informal science community.

Coordinated Activities with Other Agencies and Institutions Contributing to Objective 3.1

NIH, NSF, and universities: Community and public outreach about cancer nanotechnology is a systemic function of the NCI Alliance for Nanotechnology in Cancer. Alliance research and training centers have dedicated funds for outreach to the general, patient, and medical communities. Alliance members participate in the National Science Foundation's Nanoscale Informal Science Education Network (NISE Net) "Nano Days" either by developing their own outreach activities or by participating in activities organized in their areas. Other efforts include "Nanotechnology Town Halls" hosted by Northwestern University for a lay audience, a free one-day mini-symposium, "Nanobiomotors: Structures, Mechanisms and Clinical Implications," hosted by the University of Kentucky, which was open to the public, and "The Art of Systems Biology and Nanoscience Days for Kids and Evenings for Grownups" events hosted by the University of New Mexico. Alliance members have also developed courses and symposia on nanotechnology that are eligible for continuing medical education credits.

NIST and universities: Scientists from NIST and the University of Maryland have developed and implemented a program to introduce blind students to nanoscale science. As part of a weeklong event sponsored by the National Federation of the Blind to expose blind and visually impaired high school students from around the country to science careers, the students learned the basics of size, scale, and the metric system, and received an introduction to nanoscale concepts. Students learned through hands-on activities such as probing canes against floor models of different shapes and sizes to experience how an atomic force microscope probe senses topographic changes on a surface.

Objective 3.2 – Establish and Sustain Programs That Assist in Developing and Maintaining a Skilled Nanotechnology Workforce

Individual Agency Contributions to Objective 3.2

DOD/Air Force: Through the Research Collaboration Program, AFRL is reaching out to students nationwide and sponsoring summer internships and research at laboratories and home institutions. The focus of this work in nanomaterials and devices allows for greater exposure of these students to topics in both nanoscience and nanosafety. This initial contact with AFRL leads to ongoing relationships between students and AFRL researchers. Return interns, fellowship candidates, co-op employees, and new hires have resulted from this program, which bolsters both the overall workforce familiarity with nanotechnology issues and renews the AFRL workforce. Projects in nano-optical coatings for improved performance in imaging systems, nanomaterial biosensing systems, tunable nano-optical devices, and nano-modified fabrics for wearable electronics are among the projects that have shown promising results based on the efforts of student researchers. The experience gained by these students, both in the technical content of their work and in exposure to nanoscience laboratory practices, will augment their skills throughout their careers. [*This is also relevant to Objective 4.4.*]

DOE: By supporting fundamental and applied R&D, DOE supports students at both the graduate and undergraduate levels, as well as postdoctoral researchers, to gain expertise in nanotechnology. Additionally, the businesses supported by programs such as the SunShot Incubator and SBIR/STTR help create a skilled nanotechnology workforce. Furthermore, the Minority University Research Associates (MURA) program directly supports students working on concepts such as novel nanomaterials with optoelectronic properties tailored for PV cells, and promotes understanding of the impact of nanoparticles on thermal energy storage.

FDA: FDA training focuses on course work, including lectures by experts in science and regulatory issues concerning nanotechnology, and workshops to provide review, research, and enforcement staff laboratory training with analytical equipment used to characterize the physical and chemical properties of engineered nanomaterials relevant to FDA-regulated products. In 2015, FDA will continue to provide training for review, research, and enforcement staff, including laboratory training with analytical equipment. This type of training is especially valuable when reviewing product submissions that include data derived from such analytical equipment.

NASA: NASA continues to support the training of the next generation of scientists and engineers, including those working in nanotechnology. NASA funds graduate student research through research grants and fellowships, such as the Space Technology Research Fellows, student internships, and postdoctoral fellowships. In 2012, NASA awarded 5 new grants in nanotechnology research for a total of $3.6 million under the Experimental Program to Stimulate Competitive Research (EPSCoR) program. In late 2012, the Space Technology Research Grants Program initiated the Space Technology Early Career Faculty Awards. Of the ten grants awarded in 2013, four were in areas related nanotechnology, including the development of hierarchically ordered structures, nanocomposite membranes for water purification, and propulsion systems based on optical metamaterials.

NIH: The NCI Alliance for Nanotechnology in Cancer supports Cancer Nanotechnology Training Centers that are establishing innovative research education programs supporting the development of a cadre of investigators capable of pursuing cancer nanotechnology research. The training programs are focused on mentored laboratory-based training in multidisciplinary research projects, but each training center also develops seminars, workshops, and short courses to teach the cross-cutting skills and knowledge necessary for successful research in cancer nanotechnology. The training centers also support career development activities for their participants. The training centers have trained 125 graduate students and postdoctoral researchers in multidisciplinary research, with a focus on cross-training in medical and physical sciences and engineering. More than 800 people, ranging from undergraduates to mid-career researchers, have participated in symposia, workshops, and conferences organized or hosted by the training centers.

NIOSH: Direct training of graduate students, postdoctoral fellows, and visiting scientists in NIOSH research laboratories has been offered in the areas of toxicology, aerosol characterization, and analytical measurements. Additionally, over the past five years, NIOSH has developed and delivered several levels of training for professional industrial hygiene and

EHS practitioners in support of responsible development of the technology. Finally, through the PtD initiative, NIOSH has worked with ABET-accredited schools of engineering to introduce safety-by-design principles into engineering curricula as an additional means to develop EHS skills in the material and process designers of the future.

NSF: NSF supports education-related activities such as development of educational materials for schools, curriculum development for nanoscience and engineering, development of new teaching tools, undergraduate programs, technical training, and public outreach. Two networks for formal and informal nanotechnology education (led by Pennsylvania State University and the Museum of Science, Boston) with national outreach will continue to be supported.

USDA/NIFA: NIFA's Higher Education programs have awarded competitive grants to universities to develop nanotechnology curricula for undergraduate and graduate students in agriculture and food science and technology including food engineering, food security, food safety, bioenergy, and sustainability. For example, one project has developed courses and taught more than 600 Hispanic students in laboratory skills for plant-pathogen interaction and food safety.

Coordinated Activities with Other Agencies and Institutions Contributing to Objective 3.2

DOE and NSF: EERE-SETP and NSF co-sponsor the QESST Engineering Research Center, which has a primary goal of creating educational resources and a skilled workforce in solar technology.

NIST and NSF: NIST (CNST) and NSF (Advanced Technological Education program) are conducting a joint program that provides extended internships at NIST for community college students being trained in semiconductor manufacturing technology. The program is providing students with hands-on, practical experience in nanofabrication, processing, characterization, and tool maintenance in areas specifically targeted to meet the needs of U.S. manufacturers for skilled technicians.

The NIST/NSF jointly-funded Summer Undergraduate Research Fellowship (SURF) program provides nanotechnology research opportunities throughout NIST's laboratories and user facilities at both the Gaithersburg, MD, and Boulder, CO, campuses; in 2013 over 30 SURF students participated in nanotechnology-related projects.

NIST and industry: NIST and prominent nanotechnology companies jointly sponsored and developed an educational workshop, "Microscopy for STEM Educators," at the Scanning Microscopies 2013: Advanced Microscopy Technologies for Defense, Homeland Security, Forensic, Life, Environmental, and Industrial Sciences Conference. This workshop included a hands-on session with a commercial tabletop scanning electron microscope and an atomic force microscope.

Objective 3.3 – Provide, Facilitate the Sharing of, And Sustain the Physical R&D Infrastructure, Notably User Facilities and Cooperative Research Centers

Individual Agency Contributions to Objective 3.3

DOD/Navy: Understanding the mechanical response of nanostructured materials is key to the development of a new generation of structural materials with unprecedented properties. With the advent of new tools for characterizing mechanical response at the nanoscale level, investigations into the relationships between properties and multiple microstructural scales ranging from the nanoscale to continuum are now possible. Through Defense University Research Instrumentation Program investments, e.g., *in situ* nanoindentation in transmission electron microscopy, and cooperative agreements with the Brookhaven National Synchrotron Light Source (for time-resolved energy dispersive x-ray diffraction), university researchers are carrying out direct measurements on the evolution of microstructures and properties of nanostructured metals and ceramics during processing and deformation. This has led to the development of nanostructured metals with simultaneous high strength and high ductility, such as aluminum alloy 7075 with greater than 1 gigapascal tensile strength and 5% uniform elongation and fully dense nanostructured non-oxide ceramics, including boron carbide (B4C) for armor.

DOE: The Office of Science's five Nanoscale Science Research Centers (NSRCs) are housed in custom-designed laboratory buildings near one or more other major DOE facilities for x-ray, neutron, or electron scattering, which complement and leverage the capabilities of the NSRCs. These laboratories contain cleanrooms, nanofabrication resources, one-of-a-kind signature instruments, and other instruments not generally available except at major user facilities. These facilities are routinely made available to the academic, industrial, and nonprofit research communities during normal working hours. The NSRCs provide training for graduate students and postdoctoral associates in interdisciplinary nanoscale science, engineering, and technology research.

FDA: In 2009, FDA established two state-of-the-art nanomaterial characterization facilities, with one located on the White Oak Campus in Silver Spring, Maryland, and the other located at the National Center for Toxicological Research (NCTR) in Jefferson, Arkansas. These facilities are designed to support FDA research scientists by providing the equipment and knowledge to characterize nanomaterials. These facilities also serve to provide laboratory training for FDA review and enforcement staff. In 2015, FDA will continue to maintain these facilities and has plans to add additional equipment.

NIH: The NCI Alliance for Nanotechnology in Cancer, the NHLBI Programs of Excellence in Nanotechnology, and the NIH Roadmap for Nanomedicine Centers all follow a cooperative research model. [*See Objective 1.1.*]

NIST: NIST continues to sustain and update the capabilities and capacity in the CNST, providing industry, academia, NIST, and other Government agencies with rapid access to world-class nanoscale measurement and fabrication methods and technology. Access is provided in two ways. First, in the NanoFab, users access a commercial state-of-the-art tool set at economical hourly rates, supported by a dedicated, full-time technical support staff. Second, in the NanoLab, users access the next generation of tools and processes through collaboration with the multidisciplinary research staff. The NanoFab continues to enhance its capabilities and capacity, with notable upcoming additions, including increased capacity for electron beam lithography; multichamber, load-locked, cassette-to-cassette sputter deposition; and deep silicon etching. New capabilities in the NanoLab include methods and instrumentation for *in situ* characterization of interfaces and of nanodevices functioning in realistic operating environments, and innovative measurement methods that combine infrared spectroscopy and atomic force microscopy to determine chemical composition with nanoscale resolution.

The NIST Center for Neutron Research provides users with access to a broad range of world-class neutron scattering tools for characterizing the atomic and nanometer-scale structure and dynamics of materials.

Coordinated Activities with Other Agencies and Institutions Contributing to Objective 3.3

NIST, industry, and universities: CNST is collaborating with industry and universities to develop and make available to users new nanoscale measurement and fabrication methods. Highlights include the following new capabilities:

- In collaboration with a nanotechnology startup company, NIST has developed a new microscopy technique that uses quantum dots to allow viewing of surface and subsurface features potentially as small as 10 nanometers in size. The technique, which combines attributes of both optical and electron microscopy, has broad applications for studying a wide range of materials and systems.
- In collaboration with the University of Maryland–College Park, Syracuse University, and the University of British Columbia, NIST scientists have created an innovative, easy-to-fabricate flat lens based on negative refractive index of metamaterials, with applications in improved photolithography, nanoscale manipulation and manufacturing, and high-resolution three-dimensional imaging.

Goal 4: Support Responsible Development of Nanotechnology

Realizing the potential benefits of nanotechnology for human, social, and economic well-being, and for the environment, requires that responsible development of nanotechnology—assessment and management of potential risks—be integrated into all aspects of the field, from world-class R&D (Goal 1) to commercialization of nanotechnology-enabled products (Goal 2). Responsible development is a fundamental component of all three objectives in Goal 3. Research in support of Goal 4 addresses the recognized issues and opportunities surrounding the protection of humans and the environment, which are shared by many stakeholder groups.

In 2011, the NNI developed, with input from stakeholders, a nanotechnology-related environmental, health, and safety (nanoEHS) research strategy with a broad, multi-agency perspective.[33] That document fully supports the Goal 4 objectives and details specific research needs in six interrelated and synergistic nanoEHS areas: (1) a *nanomaterial measurement infrastructure* coupled with (2) *predictive modeling and informatics* that provides accurate and reproducible data on (3) *human exposure*, (4) *human health*, and (5) *the environment* essential for science-based *(6) risk assessment and management* of engineered nanomaterials and nanotechnology-enabled products. The NNI agencies, individually and collaboratively, will continue to provide information on progress toward addressing these research needs.

Consideration of lifecycle issues is a key component of all four objectives described below. Advances in these objectives require coordinated efforts involving multidisciplinary, multistakeholder national and international teams.

Objective 4.1 – Support the Creation of a Comprehensive Knowledge Base for Evaluation of the Potential Risks and Benefits of Nanotechnology to the Environment and to Human Health and Safety

Individual Agency Contributions to Objective 4.1

DOD/Air Force: Using core funding, AFRL scientists are accomplishing a multitude of nanotechnology research projects geared towards investigating the biological interaction of engineered nanomaterials (ENMs), including potential toxicity arising from resultant physicochemical and structural properties. This research will facilitate a better understanding of the nano–bio interaction mechanisms and provide in-depth analysis of its corresponding effects and aid researchers in devising appropriate ways to monitor exposure levels and develop control strategies to enhance Air Force force-protection efforts. While the Air Force is working to minimize issues concerning the production, handling, and disposal of nanomaterials as they relate to future mission requirements, a significant knowledge gap remains with respect to the human and ecological health implications of increasing nanomaterial usage. It is critical to understand the transport and transformation of nanomaterials channeled through the environment and the human body, including their significant avenues of access and potentially adverse effects. Two key areas of immediate military relevance include propulsion and munitions systems, which employ nanomaterials not only in reducing sensitivity to accidental initiation during storage and delivery, but in generating more energetic propulsion/explosions and ensuring long-term storage stability. Research in the Human Effectiveness Directorate focuses on understanding the toxicological properties of ENMs based on size, charge, shape, and functionalization.

DOD/Army: Understanding of potential nanoparticle release, fate, transport, and toxicity is required to understand and proactively manage risks to the environment and human health. The Army is developing tools, methods, and approaches to provide accurate and reliable information and data supporting risk assessment and management. These studies focus on the life cycle of nanotechnologies developed within the Army Research Laboratory, the Institute for Soldier Nanotechnologies, academic partners, and industry. For example, the Army, through its Engineer Research and Development Center (ERDC) and Armament Research, Development and Engineering Center (ARDEC), has begun a comprehensive program to understand the environmental consequences of nanomaterials. ARDEC will be providing ERDC with the necessary nanomaterials to support environmental testing and will help develop the database for materials that might impact DOD. Through partnerships with EPA, CPSC, NIST, and NIOSH, results of these studies will be used to support safe development of the technologies, enable more rapid fielding and commercialization, and meet regulatory requirements.

EPA: EPA is evaluating the chemical and physical properties of ENMs that influence their potential for release into the environment, fate, transport, transformation, potential for exposure to humans or sensitive ecological species, and potential for adverse effects on human health or the environment. Potential for nanomaterial releases will be evaluated across the product life cycle including manufacturing, product use, and end-of-life disposal. Identification and characterization of the role that key chemical and physical factors play in the behavior of ENMs will enable the development and parameterization of predictive models that can be used to differentiate materials and applications that may pose a higher probability of risk from those that are expected to have little impact. The predictive models can be used to direct testing and evaluation resources where there is likely to be greatest risk. In addition, this information can be used to inform the design and development of safer material applications and products.

FDA: In 2010, FDA launched its Advancing Regulatory Science Initiative, which is dedicated to the science of developing new tools, standards, and approaches to assess the safety, efficacy, quality, and performance of FDA-regulated products. One of the FDA's priority areas in this Initiative was to ensure FDA readiness to evaluate innovative emerging technologies. In 2013, FDA published its Nanotechnology Regulatory Science Research Plan, which established research categories that are linked to regulatory needs in assessing nanotechnology-enabled products. These categories include methods for physicochemical characterization, developing and evaluating *in vitro* and *in vivo* assays and models, developing and synthesizing risk characterization information, developing class-based approaches to risk assessment, and improving risk communication. Research in these categories is being coordinated through the Collaborative Opportunities for Research Excellence in Sciences (CORES) program, which is an intramural effort where projects are selected by internal and external peer review. FDA will continue the CORES program in 2015.

NIH/NIEHS: The major focus of the National Institute of Environmental Health Sciences nanotechnology research effort is broadly associated with this goal. Gaining knowledge on fundamental interaction of ENMs with biological matrices as dictated by their physicochemical properties is critical in assessing potential health risks associated with accidental or incidental exposure to nanomaterials or nanotechnology-enabled products [*see also Objectives 1.4 and 2.2*]. NIEHS investments address all of the six research needs identified in the NNI nanoEHS research strategy released in October 2011. Current NIEHS nanoEHS investments specifically focus on research needs 2, 3, 4, and 5 through the efforts of a consortium of research centers, the NIEHS Centers for Nanotechnology Health Implications Research (NCNHIR), established in 2010. These centers are evaluating the toxicity of dozens of ENMs with diverse physical and chemical properties. The ENMs being characterized by these centers represent a majority of the ENMs commercially produced in large quantities. The toxicology data from these investigations are utilized in developing computational predictive models.

NIEHS is currently developing plans to support research efforts in two areas: development of a comprehensive understanding of toxic effects of ENMs and the development of tools to assess exposure in diverse media. These efforts will focus on expanding efforts to assess the toxicity of ENMs from multiple routes of exposure. This research will also assess the potential contributions of susceptibility factors in human health effects assessment; such factors include underlying disease, genetics, age, and gender. Other new NIEHS initiatives will support research efforts to develop tools to assess human exposure to nanomaterials in different media (air and water).

NIOSH: NIOSH continues its efforts to develop more complete hazard and safety assessments using key classes of ENMs: carbon nanotubes; metal oxides; silver; the nanowire forms of silver, silica, and titania; graphene and graphene oxide; and cellulose nanocrystals and nanofibers. New research results demonstrate the effectiveness of a "safer by design" strategy by establishing decreases in adverse biologic activity resulting from modifying surface functionality. In 2015, NIOSH will move forward with development and dissemination of principles outlined in the 2012 NIOSH-sponsored Safe Nano Design workshop.[34] Key areas to continue to develop include: (1) demonstrating change in nanomaterial toxicology through material design changes and (2) partnering with private companies to develop case studies on effective facility and process design. Impact in this key program area has already been realized through NIOSH partnering with private sector companies to conduct on-site research that will result in the development and implementation of effective risk management practices that will ultimately result in safer, more efficient nanomanufacturing processes.

NIST: NIST is developing measurement tools to determine the physicochemical and human- and ecotoxicological properties of key ENMs in relevant media (e.g., air, water, and biological matrices); to characterize transformations, transport, and fate of ENMs in such media; and to evaluate exposures to airborne ENMs. NIST is also developing transferable methods to detect the presence of ENMs in nanotechnology-enabled products and to assess life-cycle releases of ENMs from such products. Key highlights of work to be conducted in 2014 and planned for 2015 include the following:

- Make available the world's first silver nanoparticle (75 nanometer and 10 nanometer diameter) and fluorescent silicon nanoparticle (3 nanometer diameter) reference materials.
- Complete comparative studies of various methods to detect the presence of silver nanoparticles in textiles and carbon nanotubes in polymer composites, including detection limits for each method.
- Design *in vitro* toxicology assays to enable evaluation of nine quality metrics that describe assay performance; identify sources of assay variability and demonstrate robustness through pilot interlaboratory testing.
- Develop methods that utilize ^{14}C-labeled CNTs to track the location and fate of CNTs in biological and environmental systems.

NRC: The Nuclear Regulatory Commission (NRC) is an independent agency created by Congress. The mission of the NRC is to license and regulate the Nation's civilian use of byproduct, source, and special nuclear materials to ensure adequate protection of the public health and safety, promote the common defense and security, and to protect the environment. As a regulatory agency, the NRC does not typically sponsor fundamental research or product development. Rather the NRC is focused, in part, on confirmatory research to verify the safe application of new technologies in the civilian nuclear industry. Currently the agency's focus with nanotechnology is to monitor developments that might be applied within the nuclear industry in order that the NRC can carry out its oversight role.

USDA/NIFA: The NIFA nanotechnology programs have been and will continue supporting the EHS research targets that are most relevant to agricultural production and food applications. For example, the NIFA Agriculture and Food Research Initiative 2014 RFA guidance states: "To ensure responsible development and deployment of nanotechnology and reap the benefits, applications should consider incorporating proper risk assessment studies as appropriate. These may include characterization of hazards and exposure levels, transport and fate of nanoparticles or nanomaterials in crops, soils (and soil biota), and livestock. This may also include animal feed formulations and processes that utilize novel materials or develop new nanostructured materials or nanoparticles that are bio-persistent in digestive pathways." In addition, research projects are sought under the AFRI Foundational Program's Improving

Food Safety topic area, with objectives to identify and characterize engineered nanoparticles and to elucidate the physical and/or molecular mechanisms that allow engineered nanoparticles to attach onto and/or internalize into fresh and fresh-cut produce, including nuts, and/or food contact surfaces associated with produce production and/or processing.

Coordinated Activities with other Agencies and Institutions Contributing to Objective 4.1

DOD/Army and Air Force: AFRL coordinates with the Army Corps of Engineers, which provides its expertise in life cycle analysis to help determine the risk of exposure to nanomaterials during different stages of development. The organizations will use predictive/computational modeling approaches to identify potential release of nanoscale particles from nanostructured materials and technologies.

DOS, NNI Agencies, and the European Union: The Transatlantic Economic Council has identified nanotechnology as a priority area for increased cooperation between the European Union and the United States. In 2011, the European Commission and the NNI organized the first U.S.-EU Workshop on Bridging nanoEHS Research Efforts. At this meeting, communities of research were proposed as a platform for scientists to develop a shared repertoire of protocols and methods, to overcome research gaps and barriers, and to enhance their professional relationships. Progress towards CoR objectives was reviewed at the third US-EU workshop on Bridging nanoEHS Research Efforts, held in Arlington, VA, on December 2–3, 2013 (us-eu.org).

CPSC, DOD/Army Corps of Engineers, EPA, FDA, NIOSH, and NIST: The CPSC staff has developed several research projects through interagency agreements in 2013 and previous fiscal years that will continue through 2014. The objective of this work is to develop robust methods to characterize nanomaterials in products and understand releases and exposure potential. The results of these studies are disseminated to the public through reports, presentations at technical meetings, and publications in peer-reviewed journals.

EPA and international organizations: EPA researchers coordinate and collaborate with other nations in the international testing program being conducted under the auspices of the OECD's Working Party on Manufactured Nanomaterials.

EPA, NSF, USDA/NIFA, CPSC, United Kingdom, and OECD: EPA's Science to Achieve Results (STAR) Program has issued joint solicitations with NSF and USDA. In addition, EPA, CPSC, and the United Kingdom awarded 6 grants in 2010 under a joint international solicitation.

FDA, NNI agencies, industry, and international organizations: As needed and appropriate, FDA continues to foster and develop collaborative relationships with other Federal agencies through the NNI, as well as with regulatory agencies, international organizations, healthcare professionals, industry, consumers, and other stakeholders.

NIH/NIEHS, EPA, NSF, FDA, and NIOSH: NIEHS has been collaborating with other Federal agencies such as EPA, NSF, FDA, and NIOSH by sharing materials and expertise. As in the past, NIH/NIEHS will invite these and other interested agencies to participate in future research initiatives, including those focused on the areas identified above in the NIH/NIEHS entry for Individual Agency Contributions to Objective 4.1.

NIOSH, CPSC, and NIH/NIEHS/NTP: NIOSH continued formal collaborations with CPSC and the National Toxicology Program (NTP) to deliver research results that provide basic knowledge in two fronts. Work done with CPSC continues to characterize the exposure potential and biological responses to inhalation of select nanomaterials from consumer products in support of human exposure risk assessments. NIOSH continues its collaboration with NTP to investigate the actual worker exposure potential during manufacture and use of ENMs. NTP also collaborates with NIOSH to investigate potential hazards associated with exposure to cellulose nanomaterials. [*Also relevant to the Nanomanufacturing NSI in Objective 1.4.*]

NIST and CPSC: NIST is developing measurement protocols and data for the release of various ENMs from consumer products during incineration, mechanical degradation, and photo-induced and hydrolytic degradation; and developing electron microscopy-based measurement protocols and data for detecting multi-walled carbon nanotubes (MWCNTs) released from commercial MWCNT-polymer composite products during mechanical abrasion.

NIST, EPA, NIOSH, CPSC, OSHA, NRC-Canada, and industry: U.S. and Canadian federal agencies are working with industry on the NanoRelease Project, coordinated by the Research Foundation of the International Life Science Institute, to identify existing methods for evaluating the release of CNTs from polymer-based composites and to develop protocols based on existing methods for use in interlaboratory testing.

NIST, FDA, and NIOSH: NIST is leading the development of measurement methods and reference materials for EHS assessments of ENMs and nanotechnology-enabled products.

NSF and EPA: NSF and EPA have renewed the large interdisciplinary Centers for the Environmental Implications of Nanotechnology (CEINs) for a second five-year period. EPA and NSF issued a joint solicitation in 2012 in the area of green chemistry, and one of the four applications awarded involves nanomaterials. The awards are approximately $5 million each.

USGS, universities, and international organizations: USGS is conducting research on the influence of physicochemical parameters and processes, such as particle aggregation and dissolution, on the bioavailability of metal nanoparticles. The research includes silver, gold, and copper nanoparticles at environmentally relevant concentrations. Stable isotopic tracers are being employed in novel ways to quantify biological uptake. For example, use of isotopically modified silver nanoparticles has proven to be an effective means of assessing bioavailability at environmentally relevant exposures (from 6 nanograms per liter to 4 micrograms per liter) and characterizing the effects of dissolution. Collaborating organizations include the University of South Carolina, The Natural History Museum of the United Kingdom, the University of Birmingham, and the University of California, Davis. These research activities are planned to continue through 2015.

Objective 4.2 – Create and Employ Means for Timely Dissemination, Evaluation, and Incorporation of Relevant EHS Knowledge and Best Practices

Individual Agency Contributions to Objective 4.2

DOD: The Emerging Contaminants Program looks "over-the-horizon" to identify and assess chemicals and materials of emerging risk. The evaluation includes the assessment of nanomaterial and involves proactive risk management actions that reduce risks to human health, the environment, and business functions, often in advance of regulatory requirements. As such, the program has issued policies on Environment, Safety, and Occupational Health (ESOH) Risks from Engineered Nanomaterials that reinforce responsibilities and provide information for managing ESOH risks of ENMs in DOD research, acquisition, operations, and support.

DOD/Air Force: AFRL is evaluating potential health risks of ENMs and identifying appropriate exposure limits, contributing to timely dissemination, evaluation, and incorporation of relevant EHS knowledge and best practices.

EPA: EPA is researching how nanomaterials affect ecosystem and human health; evaluating methods for detecting, quantifying, and characterizing nanomaterials; as well as assessing, ranking and predicting nanomaterial health effects using rapid, automated chemical screening called high-throughput testing methods. EPA is also evaluating exposure to nanomaterials at multiple levels and will provide qualified and credible alternative testing methods that can be used to predict toxicity. It also is developing a database for hazard identification or structure-activity relationships and identifying unique nanomaterials properties that can be used to design or engineer safer nanomaterials. EPA will evaluate the critical features of nanomaterials that determine their behavior in the environment including fate, transport, transformation, and potential effects on human health or the ecosystem.

FDA: FDA has made significant progress in preparing to make regulatory assessments for nanotechnologyenabled products since formally joining the NNI budget crosscut in 2011. In 2015, FDA will continue the agency-wide regulatory science program in nanotechnology to support the responsible development of FDA-regulated products containing nanomaterials or otherwise involving the application of nanotechnology. The program does the following:

- Develops the tools, methods, and data to assist in reaching regulatory decisions.
- Coordinates agency staff science and research related activities.
- Enhances in-house scientific expertise and capacity.
- Enables cooperation with other regulatory agencies.
- Facilitates regulatory science research opportunities with both domestic and international partners.

The Office of the Commissioner, in partnership with the FDA Nanotechnology Task Force, facilitates communication and cooperation on nanotechnology regulatory science, both within FDA and with national and international stakeholders. The FDA Nanotechnology Task Force also provides the overall coordination of FDA's nanotechnology regulatory science efforts in the following programmatic investment areas:

- Scientific staff development and professional training.
- Laboratory and product testing capacity.
- Collaborative and interdisciplinary research related to product characterization, manufacturing, and safety and risk assessments.

Together these program investments enable FDA to address key gaps in scientific knowledge, test methods, and predictive tools needed to make regulatory assessments for products containing nanomaterials or otherwise involving the application of nanotechnology.

NIOSH: NIOSH regularly disseminates the results of research from its nanotechnology research program in the form of publication of biannual progress reports of all NIOSH Nanotechnology Research Center projects, e-newsletters, the NIOSH Science Blog, technical meeting and symposia presentations, formal peer-reviewed journal publications, and NIOSH publications. The NIOSH Workplace Safety and Health Nanotechnology Topic webpage is one of the top pages visited on the agency website.[35] In 2015, NIOSH will continue a multifaceted initiative focused on delivering knowledge and training to industrial hygiene and EHS practitioners, private sector nanomanufacturing companies, engineering students, and international research partners. NIOSH will refine delivery of a higher-level exposure measurements and assessment strategy training course for EHS professionals. To date this has been presented to the State of California; the International Occupational Hygiene Association (Kuala Lumpur); EHS professionals in Brazil in collaboration with Fundacentro; Workplace Safety and Protection Services, Canada; at three universities; and at four U.S. professional conferences. Plans for 2015 include follow-up on the effectiveness of a planned 2014 publication on Health and Safety Prevention through Design Small Business Guide for Nanomaterial Producers and Users. In 2015 NIOSH will disseminate process-based control strategies that draw from the broader 2013 NIOSH guidance document on Current Strategies for Engineering Controls in Nanomaterial Production and Downstream Handling Processes.[36]

NSF: Research on the safety of manufacturing nanoparticles is already included in four NSECs and the NNIN. Environmental implications of nanotechnology, including development of new measurement methods for nanoparticle characterization and toxicity of nanomaterials, will continue to be investigated in two dedicated multidisciplinary centers (the Centers for Environmental Implications of Nanotechnology at UCLA and Duke University). An essential element of this will be research on methods and instrumentation for nanoparticle detection, characterization, and monitoring, including interactions of nanomaterials with cellular constituents, metabolic networks, and living tissues; bioaccumulation and its effects on living systems; and the impacts of nanostructures dispersed in the environment.

Coordinated Activities with other Agencies and Institutions Contributing to Objective 4.2

CPSC and NLM: CPSC and the National Library of Medicine (NLM) signed an interagency agreement in 2012 with work continuing through 2014, the objectives of which are to develop web content that allows consumers and businesses to access nanotechnology information and to identify areas of data enhancement for existing NLM databases. Specific outcomes include the development of a series of reports written for the typical consumer that summarize health-related information. The reports will be posted on the NLM website.

DOD, NNI agencies, and nongovernmental organizations: The Emerging Contaminants Program maintains a Nanomaterials Working Group that serves as the coordinating body for nanomaterials-related environmental, safety, and health technical, policy, and legal information. The focus of the working group includes identification and recommendation of studies and research priorities in order to avoid duplication of effort, promote leveraging of resources, issuance of consistent DOD policy and guidance, advocacy of best management practices, and coordination of risk management measures. The working group and its leadership provide means to coordinate and connect with all the military departments and DOD agencies and with other agencies and stakeholders broadly, including the Sustainable Chemicals and Materials for Defense Forum, a partnership between DOD and the American Institute for Chemical Engineers. [*Also relevant to Objective 4.4.*]

EPA, FDA, NIOSH, NIST, and OECD: OECD's Working Party on Manufactured Nanomaterials is aimed at assisting countries in their efforts to assess the safety of nanomaterials and has the active participation of U.S. bodies including EPA, FDA, NIOSH, and NIST. The OECD program has focused on methods to ensure the safety of nanomaterials, including the testing of 14 commercial nanomaterials for 59 endpoints in accordance with current OECD test guidelines (the Sponsorship Programme); the development of new guidance, test guideline modifications, and alternative methods; exposure measurement and mitigation; environmental sustainability; and risk assessment and management. By sometime in 2014, several dossiers that address testing on key nanomaterials are expected to be completed, including those for carbon nanotubes and cerium oxide; these dossiers will illustrate data trends across nanomaterials, applicability of OECD protocols, and areas for new protocol and guidance development. These areas are being explored in a series of parallel workshops beginning in 2013 and continuing through 2014, including ones on environmental fate and ecotoxicity, physicochemical properties, genotoxicity, and the grouping of nanomaterials. At least four OECD guidelines are under consideration for modification in the 2013–2014 timeframe.

EPA, NSF, USDA, CPSC, and international organizations: EPA's STAR Program has issued joint solicitations with NSF and USDA. In addition, EPA, CPSC, and the United Kingdom awarded six grants in 2010 under a joint international solicitation. EPA researchers also coordinate and collaborate with other nations in the international testing program being conducted under the auspices of the OECD's Working Party on Manufactured Nanomaterials [*see above for details*].

NSF, EPA, USDA, and the European Union: NSF will support research primarily directed at EHS implications and methods for reducing the prospective risks of nanotechnology development. NSF, EPA, USDA, and the EU will continue collaboration on development of a joint solicitation for nanotechnology EHS. A focus will be on implications of the next generation of nanotechnology products and productive processes, as well as public participation in nanotechnology-related activities. Research on both implications and applications of nanotechnology will address the sources of nanoparticles and nanostructured materials in the environment (in air, water, soil, biosystems, and working environments), as well as the nonclinical biological implications.

Objective 4.3 – Develop the National Capacity to Identify, Define, and Responsibly Address Concepts and Challenges Specific to the Ethical, Legal, and Societal Implications (ELSI) of Nanotechnology

Individual Agency Contributions to Objective 4.3

NIOSH: NIOSH is working to foster the responsible development of nanotechnology and the realization of its societal and commercial benefits. Workers are among the first in society to have potential exposures to any new material, including nanomaterials, and thus the protection of the worker can be seen as the core of responsible development. Occupational safety and health criteria for responsible development of nanotechnology should be developed at the societal level (by government agencies, trade and professional associations, unions, NGOs, and insurers) first, and then promoted for use at the business enterprise level (employers, suppliers, and business customers).

NSF: NSF supports research directed at identifying and quantifying the broad implications of nanotechnology for society, including social, economic, workforce, educational, ethical, and legal implications. The application of nanoscale technologies will stimulate far-reaching changes in the design, production, and use of many goods and services.

NSF also supports a project to embed humanists and social scientists in laboratories for greater collaboration in nanoscience around the world, providing a model for future integration of ethicists and social scientists into nanotechnology R&D laboratories broadly.

The Centers for Nanotechnology in Society at UCSB and ASU are among the very few social science centers in the world looking at nanotechnology ELSI issues and operating on a large scale to involve multiple institutions and multiple disciplines. The overarching goal of both centers is to foster the integration of social scientific research with nanoscale science and engineering to promote socially responsible innovations. CNS-UCSB does so by focusing on perceptions of risks and benefits. The full sense of perception is considered including beliefs, values, and ideologies. Risk perception is not regarded as a problem to be handled; rather, it needs to be understood ecologically, constituted by multiple stakeholder groups having multifaceted, complex, and evolving relationships. CNS-ASU does so by focusing on the unifying framework of anticipatory governance, which involves promoting multiple stakeholders in deliberation about plausible future technological developments.

The two centers promote increased awareness of societal issues within the nanoscale science and engineering community. They continue to inform policy via presentations to key state, national, and international regulators and policymakers. They also continue to engage in various public outreach engagements via media interviews, speaker series, NanoDays, web-based materials, and public presentations. CNS-ASU also studies the ways in which nanoscale science and engineering (NSE) contributes to increasing or decreasing social and economic inequalities in different national contexts, and it explores ways to ensure that NSE can contribute to equity, equality, and responsibility as public values. Over the last year, it has shifted its work to analyzing the unequal conditions and consequences of emerging nanotechnology applications in developed and developing countries. Because of limited funds, the decision was made to focus these efforts on two country case studies: the United States and South Africa. Asking these questions in the United States is an obvious responsibility of the center.

USDA/NIFA: NIFA's AFRI nanotechnology program solicits research proposals on assessment and analysis of the perceptions and social acceptance of nanotechnology and nanotechnology-based food or non-food products by the public and agriculture and food stakeholders, using appropriate social science approaches.

Coordinated Activities with other Agencies and Institutions Contributing to Objective 4.3

DOD/Air Force and EPA: AFRL has contributed substantial comments and assessments to EPA on rule development for nanomaterial reporting and information-gathering in relation to health effects, technology research environment, and impact on manufacturing development.

Objective 4.4 – Incorporate Sustainability in the Responsible Development of Nanotechnology

Individual Agency Contributions to Objective 4.4

DOD/Air Force: AFRL is providing important leadership in efforts to address this issue. AFRL scientists are conducting focused research to establish the possible effects of nanoparticle exposure on biological systems.

DOD/Army: ARL is conducting a multiscale research program to investigate low-cost, high-performance reinforcement of transparent composites using bio-derived cellulose-based nanomaterials. Target applications include lightweight films and composites for use in ballistic protection and structural applications. In particular, ARL is addressing interlayer failure in laminates that may occur through mechanical stress or by environmental factors such as heating or moisture. These improvements to the interlayer, as well as reinforcement to bulk polymers, will enable elimination of a second complementary polymer layer and more importantly elimination of the interlayer, which is often the weakest link. Target polymers representing different reinforcement needs include poly(methyl methacrylate), a high-modulus but brittle thermoplastic in need of toughening; polycarbonate, an impact-resistant thermoplastic in need of improved modulus; and a range of epoxy-based thermosetting resins. ARL is also examining a new method of transforming plastic waste materials into value-added nano-fibrous products with specific focus on water filtration membranes. Significant accomplishments include fabricated nanofibers from expanded polystyrene packaging and polyethylene terephthalate water bottles using solvent-based electrospinning.

EPA: EPA is developing predictive models pertaining to the product life cycle, environmental transport, transformation, fate, and exposure to humans or ecological populations. Current efforts are focused on evaluating existing models for applicability to nanomaterials and on identifying critical parameters that need to be specified for application to ENMs. In addition, EPA is developing approaches to evaluate effects of nanomaterials on human health or the environment based on tiered evaluations, fit-for-purpose predictive models, and structure-activity relationships among major categories of engineered materials such as metals, metal oxides, or multiwalled carbon nanotubes.

NIOSH: NIOSH is working with industry to implement effective controls and handling practices to significantly reduce worker exposure. NIOSH is also collaborating with nanomaterial manufacturers to evaluate the bioactivity of nanoparticles produced by these companies and to evaluate the effectiveness of surface modification in reducing this bioactivity.

NSF: NSF supports several programs under this objective, including the NSF-wide initiative for Science, Engineering, and Education for Sustainability (SEES), and the establishment and development of the Sustainable Nanotechnology Organization (SNO) since 2012.

Coordinated Activities with other Agencies and Institutions Contributing to Objective 4.4

DOD/Air Force, NNI agencies, universities, and industry: To examine potential toxicity, health, and environmental issues in the DOD community, Air Force personnel organized and led the first Air Force Workshop on Biological Interactions of Engineered Nanomaterials: Environmental, Safety and Health Issues of Military Concern (www.denix.osd.mil/cmrmd/upload/Nano-Flyer-Issue-1.pdf). This workshop brought together participants from industry, academia, and Government agencies, including EPA and NIOSH, to communicate and discuss current controls associated with the use of nanomaterials. AFRL scientists participate in the DOD Nanomaterials Environment Safety and Occupational Health (ESOH) Working Group. In coordination with the DOD Emerging Contaminants Governance Board, this working group provides technical, policy, and legal information relating to the safety and health issues associated with ENMs.

DOD/Army, NNI agencies, and universities: In research on the generation of bio-derived, easily processed, potentially low-cost and scalable transparent laminates using widely available cellulosic feedstock for increased sustainability, ARL has coordinated its activities with leaders in nanocellulose preparation and processing, especially focusing on critical challenges in developing optimized nanocellulose-based composites. ARL collaborates with the USDA Forest Products Laboratory, which supplies material and aids in further understanding the effects of nanocellulose surface chemistry and microstructure. ARL also partners with applied Army engineering laboratories to test and evaluate new materials. Coordinated activities also include NIST, Purdue University, the University of Maine, and the University of Wisconsin–Madison.

EPA and CPSC: The EPA and CPSC have completed a joint case study that evaluated nanoscale silver used as a disinfectant spray product. They are also collaborating in a nationwide effort to conduct innovative research to assess the effects of copper nanomaterials used to preserve wood for building decks and fences.

5. External Reviews of the NNI

The 21st Century Nanotechnology R&D Act (Public Law 108-153) calls for periodic external reviews of the NNI by the National Research Council (NRC) of the National Academies and by the National Nanotechnology Advisory Panel (NNAP). The NNAP released its last report in 2012, which was addressed in the *NNI Supplement to the President's 2014 Budget*.[37] The National Research Council issued two reports in 2013.

The first NRC report, one of those specifically required by P.L. 108-153 and entitled *Triennial Review of the National Nanotechnology Initiative*, was published in April 2013.[38] In this report, the National Research Council commends the NSET Subcommittee, the NNI agencies, and the NNCO "for their work and progress in coordinating such a diverse multiagency program" (page 14). The report also recognizes that the NNI has been "a leader among interagency initiatives in many ways" (page 14). For the NNI to continue to serve the nanotechnology community and the Nation, the National Research Council offered 20 specific recommendations in this report, organized under five crosscutting themes: (1) improve information gathering and communication at the project level; (2) develop and implement interagency plans for focused sub-areas of the NNI, particularly signature initiatives and working groups; (3) improve the NNI website, providing information specifically targeted at different stakeholders; (4) exploit advances in data collection and analysis technologies to track progress; and (5) identify, share, and implement best practices, particularly with respect to technology transfer and commercialization. The NNI agencies are assessing these recommendations and how they may be applied towards the implementation of the goals and objectives identified in the 2014 NNI Strategic Plan.

The second NRC report, *Research Progress on Environmental, Health, and Safety Aspects of Engineered Nanomaterials*, was published in September 2013.[39] It describes a set of indicators that were developed to track progress in four high-priority research areas (previously described in the 2012 National Research Council EHS Strategy[40]) and reviews the progress made through interagency coordination, stakeholder engagement, public–private partnerships, and managing potential conflict of interest. Indicators were classified "green" for new activities or expected sustained progress, "yellow" for moderate or mixed progress, or "red" for minimal activity and few anticipated changes. The report classified just one indicator as green in the "*development of methods for detecting, characterizing, tracking, and monitoring nanomaterial transformations in simple, well-characterized media.*" All other indicators for both research and implementation progress were yellow or red. However, the report notes that the interval between the NRC's 2012 EHS Strategy and its 2013 report on EHS progress was too short for substantial new research programs to be put in place and produce results. While the report suggests a change in the management structure of the NNI, the NNI member agencies and NNCO personnel view the current federated system as offering an optimal level of integration and coordination among participating agencies by connecting their mission-specific activities. Further, they view the Office of Science and Technology Policy (OSTP) and the Emerging Technologies Interagency Policy Coordination Committee (ETIPC) as providing useful guidance and expert leadership to the NNI member agencies and the NNCO.

APPENDIX A. SUPPLEMENTARY INFORMATION ON NANOTECHNOLOGY INVESTMENTS BY THE DEPARTMENT OF DEFENSE[41]

Supplementary Information on DOD Research Activities

Air Force: Nanomaterials and nanodevices continue to offer new application potential by improving response, efficiency, and maintainability in materials systems. By identifying novel materials with properties that have specific Air Force relevance, AFRL has targeted research programs to investigate potentially revolutionary advances in performance such as through 2-D electronic materials, novel active ceramic materials for laser sources, and high-performance nanotube fibers. Notable advances combining self-assembled nanosystems and optical nanocomposites through collaborative industrial-governmentacademic research teams have received significant attention at national and international conferences. International partnerships in nanomagnetics have led to significant progress toward nanoparticle magnetic inks that will enable additive magnetic patterning on structural components for electromagnetic applications including conformal antennas.

Army*:* The U.S. Army Engineer Research and Development Center has a multidisciplinary nanotechnology research program with emphasis on the development of computational materials modeling, simulation, and design capabilities to support the production of new classes of novel, high-performance materials for use in military engineering applications. This program has focused on developing a better understanding of the formation and influence of a materials nanostructure on its nano-, micro-, and mesoscale performance through the development of unique diagnostic capabilities, computational tools, and advanced synthesis methods that will be crucial in the development of advanced, multiscale, reinforced cementitious composites for facility protection.

The Army Research Laboratory is furthering the fundamental understanding and application of nanotechnology to enhancing weapons, protective systems, and warfighter capabilities. The challenges that impede the rapid transition of nanotechnology to applications and fielding are synthesis by design and processes suitable for application-pull. Whether the product is nanostructured materials for armor or nanoparticles for protective coatings, the pervasive technical challenge in processing and manufacturing of nanomaterials is the ability to control disbursement, distribution, stability, morphology, and microstructural design throughout the synthesis and consolidation phases. The goal of this program is to resolve the fundamental barriers to achieving the unique properties and transition paths to bring nanotechnology-enabled capabilities to the soldier. The major challenges are nucleation, grain growth, and particle agglomeration. Some research highlights include the production of bulk nanostructured tungsten for replacing depleted uranium in kinetic energy penetrator applications, development of technologies to stabilize interfaces and grain boundaries to improve the thermal stability of nanomaterials, and severe plastic deformation processing as a cost-effective means to produce nanostructured metals for armor and penetrator applications. ARL has also developed methodologies to utilize biomolecules (i.e., proteins, DNA, etc.) to achieve atomic-level control of nanoparticle synthesis. ARL has demonstrated these new bio-nanomaterials applicable in biosensing and pressure sensing for mild traumatic brain injury.

The Army-sponsored Institute for Soldier Nanotechnologies at MIT is conducting research to dramatically improve the survivability of the soldier by extending the frontiers of nanotechnology through fundamental research and transitioning promising outcomes of that work in collaboration with the Army and industry partners. This includes lightweight nanocrystalline alloys with increased strength and energy dissipation characteristics for blast and ballistic protection and fabricated nanoparticle metal alloys that combine the high strength of steel with the low weight and high ductility of aluminum.

CBDP: The Chemical and Biological Defense Program (CBDP) has made investments in new and ongoing single- and multi-investigator projects in the areas of (a) advanced drug delivery platforms, including those that have improved targeting of non-host pathogens and delivery of antibiotic and nerve agent countermeasure payloads to improve drug efficiency; (b) advanced nanotechnology-enabled sensing and diagnostics platforms, including those that enlist new understanding of common biological mechanisms of threat action as well as novel fluidics and plasmonics approaches; (c) nanostructured materials for protection, hazard mitigation, and hazard elimination capabilities; (d) nanostructured materials for control of micro- and nanoscale separation and transport; and (e) understanding and modulation of biotic/abiotic interfaces. These areas are viewed as strategically relevant to development of future nanotechnologyenabled chemical and biological defense needs such as threat sensing, point-of-care diagnostics, and threat countermeasures.

DARPA: DARPA is funding leading U.S. universities to advance nanotechnologies for creating useful material systems, devices, and architectures for future microsystems. DARPA, in collaboration with companies from the semiconductor and defense industries, including Applied Materials, Global Foundries, IBM, Intel, Micron, Raytheon, Texas Instruments, and United Technologies, established the Semiconductor Technology Advanced Research Network (STARnet). This effort builds a large multi-university research community to explore beyond current evolutionary directions to make discoveries that drive technology innovation beyond what can be imagined for electronics today. The universities are organized into six multi-university centers, each focused in a specific challenge area. Researchers in these centers are advancing R&D in physics/theory, materials, processes, circuits, and related issues (design, off-chip communications, etc.) for new devices that operate using unique information tokens due to physical phenomena that occur at the nanoscale. Technologies satisfying future demands of military and civilian systems require harnessing the quantum nature and nonlinearities of nanoscale devices to transition from classical to quantum engineering. Advances in several scientific disciplines accelerate this transition. Under DARPA sponsorship, a new state of matter was discovered—topological insulators with insulating bulk and topologically ordered metallic surfaces—resulting in remarkable properties that are promising for electronics beyond the current state of the art. Other developments include new approaches to sensing and information transduction, techniques to control quantum states, and progress in nanoscale manufacturing. DARPA's MesoDynamic Architectures (Meso) program works at the intersection of these disciplines, exploiting mesoscale and nonlinear dynamics to create technologies with exotic functionality and unprecedented performance in sensing, computation, communication, and navigation, and that are amenable to harsh environments. Accomplishments include very high-performance microelectromechanical systems (MEMS) oscillators, the first piezoelectronic transistor, an electronic biomolecular sensor detecting minimum injury concentrations of neurotoxins, and the first-ever topological insulator thermoelectric device.

DTRA*:* The Defense Threat Reduction Agency Basic and Applied Sciences program supports investigations of nanoscience that may lead to the development of nanotechnologies applicable to a number of mission needs for countering weapons of mass destruction (WMDs). Areas of interest include tamper-evident nanostructures to help with securing and monitoring WMDs, nanoelectronics for radiation-hardening and life extension of counter-WMD systems, conventional weapons-effect testing with novel diagnostics, radiation sensing systems with novel interfacial materials, and alternative methods for photon or electron multiplication. Programmatic interests are supported through current investment in exploratory research efforts in areas such as (1) engineered nanomaterials for unattended sensing of WMDs; (2) nanomaterials as fast-responding diagnostics to measure temperature, pressure, or mechanical strain within reacting energetic materials; (3) 2-D nanomaterials for development of microelectronic components with low radiation cross section; (4) radiation effects in nanophotonic resonators and waveguides; and (5) nanoparticle scintillators for radiation detection. Examples of significant research include the work at Washington State University demonstrating that novel europium (Eu)-doped yttrium oxide (Y_2O_3) and erbium (Er)/ytterbium (Yb)-doped zirconium dioxide (ZrO_2) can evaluate the thermal history of a detonation to within 50 °C and 10 second accuracy, and at Arizona State University where polypeptidebased nanomatrices were explored for their use in sensor materials and further shown to have potential spin-off medical applications for laser tissue healing.

Navy: Significant advances are being made in the understanding and manipulation of nanomaterials in a wide variety of program areas, including quantum computing, nanoelectronics, power generation, and "Lighten the Load" technologies for warfighters and combat vehicles. Researchers at the Naval Research Laboratory recently achieved a breakthrough in photonic crystal structures, which are of interest for controlling light on a microscopic scale. They can route light through photonic circuits and concentrate light into very small volumes, and can provide an interface between light and a single quantum emitter. Ultimately, such a system can be used for nanoscale lasers, low-power optical switching, and for quantum information applications. For quantum information, a photonic crystal cavity coupled to control of long-lived quantum bits (qubits) provides the essential building block of a quantum network, which can be used for quantum communication or quantum computing, including secure communications and rapid processing of large data sets. NRL has made this crucial step by developing a quantum dot spin qubit that is both long-lived and optically active, and incorporating it into an optical cavity in a photonic crystal. A universal single-qubit logic gate has been demonstrated, as well as efficient and highly frequency-tunable photon emission.

The *International Technology Roadmap for Semiconductors* (ITRS) has identified the electron's spin angular momentum as an alternate state variable for information processing to enable minimization beyond the traditional size scaling of Moore's Law. Electrical injection and detection of spin in a semiconductor using a ferromagnetic metal contact requires an intervening tunnel barrier to accommodate the large mismatch in conductivity. This results in unacceptably high contact resistances, which prevent further technological development. Scientists at NRL have demonstrated that graphene, a single layer of carbon atoms in a honeycomb lattice, can serve as a low-resistance spin-polarized tunnel barrier contact that successfully enables efficient spin injection/detection in silicon. They have demonstrated electrical generation and detection of spin accumulation in silicon above room temperature and showed that the contact resistance–area products are two to three orders of magnitude lower than those achieved with conventional oxide tunnel barriers. This approach utilizes a defect resistant, chemically inert and stable material with well-controlled thickness to achieve a low-resistance spin contact compatible with both the ferromagnetic metal and semiconductor of choice, and it introduces a new paradigm for tunnel barrier technology and significantly advances technological applications of silicon-based spintronics. These contacts were also utilized for spin injection/detection in silicon nanowires and enabled the first observation of Hanle spin precession and the first direct measurement of spin lifetimes in silicon nanowires up to room temperature.

Changes in Balance of Investments across PCAs by Military Departments and DOD Agencies

Air Force: The President's 2015 Budget request for the Air Force will continue to support research in nanotechnology. Classification of 2013 and 2014 investments highlight a focus on PCA 1 (Nanotechnology Signature Initiatives), PCA 2 (Foundational Research), PCA 3 (Nanotechnology-Enabled Applications, Devices, and Systems), and PCA 4 (Research Infrastructure and Instrumentation). An approximate comparison to 2013 levels indicates that the requests for 2015 will decrease the budget for PCAs 1 and 2 by 7% each and decrease the Air Force's investments in PCA 3 by 7%. The changes in overall funding between 2014 and 2015 have occurred largely due to the expiration of Congressional Interest Items and because of better classification of current and planned research under the revised PCAs. Requests for the 2015 programs are mainly continuations of 2014 efforts, with limited new-start proof-of-concept efforts beginning in 2015 in the areas yet to be determined. The relative balance across PCAs from 2013 to 2015 is associated with regular program turnover and transition activities and is quite stable, except in nanoelectronics, due to project completions.

Army: The President's 2015 Budget request for the Army will continue to support research in nanotechnology primarily in the areas of PCA 2 (Foundational Research), PCA 3 (Nanotechnology-Enabled Applications, Devices, and Systems), and PCA 5 (Environment, Health, and Safety). For the Army, the 2015 estimate shows an 11% overall decrease and some shifts between PCAs, especially an increase in PCA 2 and commensurate decrease in PCA 3 due to better classification of current and planned research under the new PCAs.

CBDP: The Chemical and Biological Defense Program current nanotechnology investments are aligned according to the budget activity of funding. Efforts in PCA 2 (Foundational Research) are funded by Budget Activity 1 (basic research), which is concerned with advancing scientific knowledge relevant to the CBDP mission, but without specific applications or immediate commercial objectives. Efforts in PCA 3 (Nanotechnology-Enabled Applications, Devices, and Systems) are funded by Budget Activity 2 (applied research), which is intended for gaining the knowledge necessary to determine the means to meet a recognized or specific need. The current balance of investment shows a greater focus on PCA 2, which reflects the program's historical and current emphasis on understanding and predicting phenomena at the nanoscale in order to enable improved capabilities for chemical and biological defense. The program saw an overall increase in 2013 and 2014 nanotechnology funding, reflecting the start of several projects in both PCA 2 and PCA 3. The 50% decrease in the 2015 estimate compared to 2014 reflects the exploratory nature of these new programs, as well as normal programmatic fluctuation, and should not be viewed as a cut in overall nanotechnology investment. Increased follow-on funding in 2015 is possible depending on the technical success of these new and ongoing programs, the availability of funds, and prioritization of potential new 2015 topic areas in nanotechnology according to CBDP strategy.

DARPA: The President's 2015 Budget request for DARPA will continue to support research in nanotechnology across three key PCAs: Nanotechnology Signature Initiatives (PCA 1); Foundational Research (PCA 2); and Nanotechnology-Enabled Applications, Devices, and Systems (PCA 3). Compared with 2013 levels, the estimates in 2014 increase the total budget for PCA 1 by 26%. This 26% increase reflects the following redistribution of funds within the PCA 1 Signature Initiatives: (1) a reduction of $0.1 million in PCA 1 Signature Initiative "Nanoelectronics for 2020 and Beyond" associated with the planned completion of the Atomic Scale Materials and Devices program; and (2) an increase of $4.5 million in PCA 1 Signature Initiative "Nanotechnology for Sensors and Sensors for Nanotechnology" associated with planned program funding increases in the Basic Photon Science and Enabling Quantum Technologies programs. Compared with 2013 levels, the 2014 and 2015 reduction in funding requests for PCA 2 reflect the planned completion of the Physics in Biology program. Specifically, the funding for Physics in Biology is reduced by 33% to $0.2 million in 2014, and no funding is requested in the 2015 Budget. Also, the funding for STARnet remains balanced at $2.0 million/year across 2013 to 2015. Compared with 2013 levels, the 2014 and 2015 reductions in funding requests for PCA 3 reflect the planned completion of the Meso program.

DTRA: The DTRA Basic and Applied Sciences program continues to support the NNI through targeted research programs geared toward the exploration of novel materials and processes relevant to the counter-weapons of mass destruction (C-WMD) mission space. Classification of 2013 and 2014 investments highlight a focus on PCA 2 (Foundational Research). The estimated 15% decrease in overall investment for 2014 is related to shifts in programmatic priorities for the C-WMD mission. Changes are not indicative of planned decreases in nanotechnology investments but reflect technical successes for projects, availability of R&D funding, and prioritization of thrust areas in accordance with the DTRA research strategy.

Navy: The President's 2015 Budget request for the Navy (Office of Naval Research and Naval Research Laboratory—ONR/NRL) will continue to support research in nanotechnology. Classification of 2013 and 2014 investments highlight a focus on PCA 1, especially in nanoelectronics and knowledge infrastructure NSIs, and PCA 2 (Foundational Research) with continuing investments in PCA 3 (Devices and Systems). Redistribution of investments across the PCAs has occurred due to the refinement of PCAs from 2013 to 2015. The overall investment is estimated to remain stable from 2013 to 2014 with a 26% estimated decrease in the 2015 request. This reduction is consistent with overall Department of the Navy budget realignments. The 2015 Budget request will maintain focus on the Nanotechnology Signature Initiatives; Foundational Research; and Nanotechnology-Enabled Applications, Devices, and Systems in support of defense applications.

APPENDIX B. ABBREVIATIONS AND ACRONYMS

AFOSR	Air Force Office of Scientific Research
AFRI	Agriculture and Food Research Initiative
	(USDA/NIFA)
AFRL	Air Force Research Laboratory
ARL	Army Research Laboratory
ARO	Army Research Office (DOD)
ARS	Agricultural Research Service (USDA)
BIS	Bureau of Industry and Security (DOC)
BRIDGE	The Bridging Research Interactions through collaborative Development Grants in Energy (DOE FOA)
CBDP	Chemical and Biological Defense Program
CDC	Centers for Disease Control and Prevention (DHHS)
CMOS	complementary metal-oxide semiconductor
CN	cellulose nanomaterial(s)
CNS	Centers for Nanotechnology in Society at Arizona State University and University of California, Santa Barbara (CNS-ASU and CNS-UCSB) (NSF)
CNST	Center for Nanoscale Science and Technology (DOC/NIST)
CNT	carbon nanotube
CoR	Communities of Research
CPSC	Consumer Product Safety Commission
DARPA	Defense Advanced Research Projects Agency
DHHS	Department of Health and Human Services
DHS	Department of Homeland Security
DNI	Director of National Intelligence
DOC	Department of Commerce
DOD	Department of Defense
DOE	Department of Energy
DOEd	Department of Education
DOE-SC	DOE Office of Science

DOJ	Department of Justice
DOL	Department of Labor
DOS	Department of State
DOT	Department of Transportation
DOTreas	Department of the Treasury
DTRA	Defense Threat Reduction Agency (DOD)
EERE	(Office of) Energy Efficiency and Renewable Energy (DOE)
EHS	environment(al), health, and safety
ELSI	ethical, legal, and societal implications
ENG	Engineering Directorate of NSF
ENM	engineered nanomaterial
EPA	Environmental Protection Agency
ERC	Engineering Research Centers (NSF)
ERDC	Engineer Research and Development Center
EU	European Union
FDA	Food and Drug Administration (DHHS)
FHWA	Federal Highway Administration (DOT)
FOA	Funding Opportunity Announcement
FS	Forest Service (USDA)
hBN	hexagonal boron nitride
HFCT	Hydrogen and Fuel Cells Technology program (DOE)
IC	Intelligence Community
ISN	Institute for Soldier Nanotechnologies at MIT (DOD/Army)
ISO	International Organization for Standardization
LED	light-emitting diode
Meso	Mesodynamic Architectures (program) (DOD/DARPA)
MPS	Mathematical and Physical Sciences Directorate of NSF
nanoEHS	nanotechnology environment, health, and safety (research, etc.)
NASA	National Aeronautics and Space Administration
NCI	National Cancer Institute (DHHS/NIH)
NCL	Nanotechnology Characterization Laboratory (DHHS/NIH/NCI)
NEHI	Nanotechnology Environmental and Health Implications Working Group of the NSET Subcommittee
NGO	nongovernmental organization
NHLBI	National Heart, Lung, and Blood Institute (DHHS/NIH)
NIEHS	National Institute of Environmental Health Sciences (DHHS/NIH)
NIFA	National Institute of Food and Agriculture (USDA)
NIH	National Institutes of Health (DHHS)
NIOSH	National Institute for Occupational Safety and Health (DHHS/CDC)
NIST	National Institute of Standards and Technology (DOC)
NITRD	Networking and Information Technology Research and Development Program

Appendix B. (Continued)

NKI	Nanotechnology Knowledge Infrastructure (Nanotechnology Signature Initiative)
nm	nanometer(s)
NNCO	National Nanotechnology Coordination Office
NNI	National Nanotechnology Initiative
NNIN	National Nanotechnology Infrastructure Network
NRC	Nuclear Regulatory Commission
NRC-Canada	National Research Council of Canada
NRI	Nanoelectronics Research Initiative
NRO	National Reconnaissance Office (IC/DNI)
NSE	nanoscale science and engineering
NSEC	Nanoscale Science and Engineering Centers (NSF)
NSET	Nanoscale Science, Engineering, and Technology Subcommittee of the NSTC
NSF	National Science Foundation
NSI	Nanotechnology Signature Initiative
NSRC	Nanoscale Science Research Centers (DOE)
NSTC	National Science and Technology Council
NTP	National Toxicology Program (DHHS/multiagency)
OECD	Organisation for Economic Co-operation and Development
OLED	organic light-emitting diode
OMB	Office of Management and Budget (Executive Office of the President)
ONR	Office of Naval Research
OSHA	Occupational Safety and Health Administration (DOL)
OSTP	Office of Science and Technology Policy (Executive Office of the President)
PCA	Program Component Area of the National Nanotechnology Initiative
PtD	Prevention through Design
PV	photovoltaic
QESST	Quantum Energy and Sustainable Solar Technologies (NSF/DOE Engineering Research Center)
RFA	request for applications
RFI	request for information
SBIR	Small Business Innovation Research Program
SETP	Solar Energy Technology Program (DOE/EERE)
SI	International System of Units
SSL	solid-state lighting
STAR	Science to Achieve Results (EPA)
STARnet	Semiconductor Technology Advanced Research Network (DARPA, NSF, NIST, industry)
STEM	science, technology, engineering, and mathematics

STTR	Small Business Technology Transfer Research Program
TAPPI	Technical Association of the Pulp and Paper Industry
TONIC	Translation of Nanotechnology in Cancer
USDA	U.S. Department of Agriculture
USGS	U.S. Geological Survey
USITC	U.S. International Trade Commission
USPTO	U.S. Patent and Trademark Office (DOC)
WMD(s)	weapon(s) of mass destruction

APPENDIX C. CONTACT LIST

OSTP
Dr. Altaf H. Carim
Assistant Director for Nanotechnology
Office of Science and Technology Policy
Executive Office of the President
Washington, DC 20502
T: 202-456-7825
acarim@ostp.eop.gov

OMB
Dr. Danielle Jones
Office of Management and Budget
Executive Office of the President
725 17th St., N.W.
Washington, DC 20503
T: 202-395-1741
Danielle_Y_Jones@omb.eop.gov

Dr. James Kim
Office of Management and Budget
Executive Office of the President
725 17th St., N.W.
Washington, DC 20503
T: 202-395-3085
James_H_kim@omb.eop.gov

Ms. Celinda Marsh
Office of Management and Budget
Executive Office of the President
725 17th St., N.W.
Washington, DC 20503
T: 202-395-5835
Celinda_A._Marsh@omb.eop.gov

NNCO[42]

Dr. Lloyd J. Whitman
Interim Director
National Nanotechnology
Coordination Office
4201 Wilson Blvd.
Stafford II, Suite 405
Arlington, VA 22230
T: 703-292-8626; F: 703-292-9312
lwhitman@nnco.nano.gov

CPSC

Dr. Mary Ann Danello
Associate Executive Director
Directorate for Health Sciences
Consumer Product Safety Commission
4330 East-West Highway, Suite 600
Bethesda, MD 20814
T: 301-504-7237; F: 301-504-0079
mdanello@cpsc.gov

Dr. Treye Thomas
Leader, Chemical Hazards Program
Office of Hazard Identification & Reduction
Consumer Product Safety Commission
4330 East-West Highway,
Suite 600
Bethesda, MD 20814
F: 301-987-2560; F: 301-504-0079
tthomas@cpsc.gov

DHS

Dr. Richard T. Lareau
Science and Technology Directorate
Department of Homeland Security
William J. Hughes Technical Center, Bldg. 315
Atlantic City International Airport
NJ 08405
T: 609-813-2760; F: 609-383-1973
Richard.Lareau@dhs.gov

Mr. Ron McNeal
Department of Homeland Security
Office of Research and Development
Science and Technology Directorate
1120 Vermont Ave., N.W.

Washington, DC 20005
T: 202-254-6772
ron.mcneal@hq.dhs.gov

DNI

Mr. Richard D. Ridgley
Chief Scientist, Advanced Systems and
Technology Directorate
National Reconnaissance Office
14675 Lee Road
Chantilly, VA 20151
T: 703 808-3115; F: 703 808-2646
ridgleyr@nro.mil

Mr. Matthew J. Cobert
Nanotechnologies Branch Chief
National Reconnaissance Office
14675 Lee Road
Chantilly, VA 20151
T: 703 808-3115; F: 703 808-2646
cobertma@nro.mil

DOC/BIS

Ms. Kelly Gardner
Export Policy Advisor
U.S. Department of Commerce
Bureau of Industry and Security
Office of National Security and Technology
Transfer Controls, Rm. 2600
14th St. & Constitution Ave., N.W.
Washington, DC 20230
T: 202-482-0102; F: 202-482-3345
Kelly.Gardner@bis.doc.gov

DOD

Dr. Jeffrey DePriest
New Technologies Program Manager
Counterforce Systems Division
Defense Threat Reduction Agency
8725 John J. Kingman Road, MSC 6201
Fort Belvoir, VA 22060-6201
T: 703-767-4199; F: 703-767-5535
Jeffrey.DePriest@dtra.mil

Dr. Mostafiz Chowdhury
The Warrior Injury Assessment Manikin
(WIAMan) Project Management Office
(PMO)
U.S. Army Research Laboratory
RDRL-DW
2800 Powder Mill Road
Adelphi, Maryland 20783-1145
T: 301-394-2917
Mostafiz.R.Chowdhury.civ@mail.mil

Dr. Anthony Esposito
Science & Technology Manager
Chemical & Biological Technologies
Department
Defense Threat Reduction Agency
8725 John J. Kingman Road, Stop 6201
Fort Belvoir, VA 22060-6201
T: 703-767-3366; F: 703-767-1893
anthony.esposito@dtra.mil

Dr. Akbar S. Khan
Senior Microbiologist & S&T Manager
Basic & Supporting Research Div.
Defense Threat Reduction Agency
Fort Belvoir, VA 22060-6201
T: 703-767-3391; F: 703-767-1893
Akbar.Khan@dtra.mil

Dr. Heather Meeks
Physical Scientist
Basic & Supporting Research Div.
Defense Threat Reduction Agency
Fort Belvoir, VA 22060-6201
T: 703-767-3059; F: 703-767-1893
Heather.Meeks@dtra.mil

Dr. Brian Pate
Physical Scientist
Basic & Supporting Research Div.
Defense Threat Reduction Agency
Fort Belvoir, VA 22060-6201
T: 703-767-3388; F: 703-767-1893
Brian.Pate@dtra.mil

Dr. Gernot S. Pomrenke
Program Manager - Optoelectronics, THz
and Nanotechnology
AFOSR/NE
Directorate of Physics and Electronics
Air Force Office of Scientific Research
875 North Randolph Street
Suite 325, Rm. 3112
Arlington, VA 22203-1768
T: 703-696-8426; F: 703-696-8481
gernot.pomrenke@afosr.af.mil

Dr. Lewis Sloter
Associate Director, Materials & Structures
Office of the Assistant Secretary of
Defense for Research and Engineering
AT&L (ASDR&E)
4800 Mark Center Dr.
Alexandria VA 22350-3600
T: 571-372-6511; F: 571-372-6500
Lewis.E.Sloter.civ@mail.mil

Dr. David M. Stepp
Chief, Materials Science Division
Army Research Office
AMSRD-ROE-M (Materials Science Division)
P.O. Box 12211
Research Triangle Park, NC 27709
T: 919-549-4329; F: 919-549-4399
david.m.stepp.civ@mail.mil

DOE

Dr. Harriet Kung
Director, Office of Basic Energy Sciences
Office of Science
U.S. Department of Energy
SC-22/GTN
1000 Independence Ave., S.W.
Washington, DC 20585
T: 301-903-3081; F: 301-903-6594
Harriet.Kung@science.doe.gov

Dr. George Maracas
Division of Scientific User Facilities
Office of Basic Energy Sciences
U.S. Department of Energy

SC-22.3/GTN
1000 Independence Avenue, S.W.
Washington, D.C. 20585
T: 301-903-0064; F: 301-903-1690
george.maracas@science.doe.gov

Dr. John C. Miller
Division of Chemical Sciences,
Geosciences, and Biosciences
Office of Basic Energy Sciences
U.S. Department of Energy
SC-22.1/GTN
1000 Independence Ave., S.W.
Washington, DC 20585
T: 301-903-5806; F: 301-903-4110
John.Miller@science.doe.gov

Dr. Andrew R. Schwartz
Office of Basic Energy Sciences
U.S. Department of Energy
SC-22/GTN
1000 Independence Avenue SW
Washington, DC 20585-1290
T: (301) 903-3535; F: (301) 903-9513
andrew.schwartz@science.doe.gov

Dr. Brian Valentine
Advanced Manufacturing Office
U.S. Department of Energy
Headquarters Bldg., EE-2F
1000 Independence Avenue, S.W.
Washington, D.C. 20585-0121
T: 202-586-9741; F: 202-586-9234
Brian.Valentine@ee.doe.gov

DOEd

Mr. Peirce Hammond
Senior Advisor for Special Initiatives
U.S. Department of Education
Office of Vocational and Adult Education
Room 11-106 Potomac Center Plaza
400 Maryland Avenue, SW
Washington, DC 20202-7100
T: 202-245-6056
Peirce.Hammond@ed.gov

DOJ

Mr. Joseph Heaps
Policy Advisor, Communications and Radio Frequency Spectrum Issues
National Institute of Justice
Office of Science and Technology
810 7th Street, NW, Rm. 7206
Washington, DC 20531
T: 202-305-1554; F: 202-305-9091
Joseph.Heaps@usdoj.gov

DOL/OSHA

Dr. Janet Carter
Senior Health Scientist
Department of Labor
Occupational Safety and Health Administration
Room N3718
200 Constitution Ave., NW
Washington, DC 20210
T: 513-684-6030; F: 513-684-2630
Carter.Janet@dol.gov

DOS

Mr. Kenneth Hodgkins
Director
Office of Space & Advanced Technology
U.S. Department of State
OES/SAT, SA-23
1990 K Street, NW, Suite 410
Washington, DC 20006
T: 202-663-2398; F: 202-663-2402
HodgkinsKD@state.gov

Dr. Chris Cannizzaro
Physical Science Officer
Office of Space & Advanced Technology
U.S. Department of State
OES/SAT, SA-23
1990 K Street, NW, Suite 410
Washington, DC 20006
T: 202-663-2390; F: 202-663-2402
CannizzaroCM@state.gov

DOT

Mr. Peter Chipman
Office of the Assistant Secretary for Research and Technology
Department of Transportation
1200 New Jersey Ave., S.E.
Washington, DC 20590
T: 202-366-0934
Peter.Chipman@dot.gov

Dr. Jonathan Porter
Chief Scientist
Office of Research, Development, and Technology
Federal Highway Administration
U.S. Department of Transportation
Turner-Fairbank Highway Research Center
6300 Georgetown Pike
McLean, VA 22101-2200
T: 202-493-3038; F: 202-493-3417
jonathan.porter@dot.gov

DOTreas

Dr. John F. Bobalek
Program Manager
Office of Research and Technical Support
Bureau of Engraving and Printing
Rm. 201-29A
14th and C Streets, SW
Washington, DC 20228-0001
T: 202-874-2109; F: 202-927-7415
John.Bobalek@bep.treas.gov

DOC/EDA

Mr. Thomas Guevara
Deputy Assistant Secretary for Regional Affairs
U.S. Department of Commerce
1401 Constitution Avenue, NW
Suite 71014
Washington, DC 20230
T: 202-482-5891, F: 202-273-4781
tguevara@eda.doc.gov

EPA

Dr. Tina Bahadori
National Program Director
Chemical Safety for Sustainability
Office of Research and Development
Mail Code 8101R
Environmental Protection Agency
1200 Pennsylvania Avenue, NW
Washington, DC 20460
T: 202-564-0428
Bahadori.Tina@epa.gov

Dr. William Boyes
Interim Associate National Program Director
Chemical Safety and Sustainability
Office of Research and Development
Mail Code B105-05
Environmental Protection Agency
109 T.W. Alexander Drive
Research Triangle Park, NC 27711
T: 919-541-7538; F: 919-541-4849
boyes.william@epa.gov

Dr. Nora F. Savage
National Center for Environmental Research
Office of Research and Development
Mail Code 8722F
Environmental Protection Agency
1200 Pennsylvania Avenue, NW
Washington, DC 20460
T: 202-347-8104; F: 202-347-8142
Savage.Nora@epamail.epa.gov

Dr. Philip Sayre
Associate Director
Risk Assessment Division
Office of Pollution Prevention and Toxics
Environmental Protection Agency
1200 Pennsylvania Ave., NW
Mail Code 7403M
Washington, DC 20460
T: 202-564-7673; F: 202-564-7450
sayre.phil@epa.gov

FDA
Dr. Jed Costanza
Microbiologist
Office of the Chief Scientist
Food and Drug Administration
10903 New Hampshire Avenue
WO Bldg 32 Room 4153
Silver Spring, MD 20993
T: 240-402-4120; F: 301-847-8614
jed.costanza@fda.hhs.gov

NASA
Dr. Michael A. Meador
Nanotechnology Project Manager, Game Changing Technologies
National Aeronautics and Space Administration
Glenn Research Center
21000 Brookpark Road
Cleveland, OH 44135
T: 216-433-9518; F: 216-977-7132
Michael.A.Meador@nasa.gov

Dr. Meyya Meyyappan
National Aeronautics and Space Administration
Ames Research Center
MS 229-3
Moffett Field, CA 94035
T: 650-604-2616; F: 650-604-5244
m.meyyappan@nasa.gov

NIH
Dr. Lori Henderson
Program Director
Clinical Trials Branch/Cancer Imaging Program
Division of Cancer Treatment and Diagnosis
National Cancer Institute
6130 Executive Blvd.
Rockville, MD 20892
T: 240- 276-5930; F: 240-276-7890
hendersonlori@mail.nih.gov

NIH/OD

Dr. Nancy E. Miller
Senior Science Policy Analyst
Office of Science Policy & Planning
Office of the Director
National Institutes of Health
1 Center Drive
Bldg 1, Rm. 218
Bethesda, MD 20892
T: 301-594-7742; F: 301-402-0280
Nancy.Miller1@nih.gov

NIH/FIC

Mr. George Herrfurth
Multilateral Affairs Coordinator
Division of International Relations
Fogarty International Center
National Institutes of Health
Building 31, Rm. B2C11
Bethesda, MD 20892-2220
T: 301-402-1615; F: 301-480-3414
herrfurg@mail.nih.gov

NIH/NCI

Dr. Piotr Grodzinski
Program Director, Nanotechnology in Cancer
Office of Technology & Industrial Relations
Office of the Director
National Cancer Institute
National Institutes of Health
31 Center Drive, 10A52
Bethesda, MD. 20892
T: 301-496-1550; F: 301-496-7807
grodzinp@mail.nih.gov

Dr. Scott McNeil
Director, Nanotechnology
Characterization Laboratory
National Cancer Institute at Frederick
National Institutes of Health
P.O. Box B, Building 469
1050 Boyles Street
Frederick, MD 21702-1201
T: 301-846-6939; F: 301-846-6399
mcneils@ncifcrf.gov

NIH/NHLBI
Dr. Denis Buxton
Basic and Early Translational Research
Program, DCVS
National Heart, Lung, and Blood Institute
National Institutes of Health
6701 Rockledge Drive, Rm. 8216
Bethesda, MD 20817
T: 301-435-0513; F: 301-480-1454
BuxtonD@nhlbi.nih.gov

NIH/NIBIB
Dr. Steven Zullo
Scientific Review Officer
Office of Scientific Review
National Institute of Biomedical Imaging
and Bioengineering
Building Democracy II-952
6707 Democracy Blvd
Bethesda, MD 20892
T: 240-271-9007; F: 301-480-1675
Steven.Zullo@nih.hhs.gov

NIH/NIEHS
Dr. John R. Bucher
Associate Director
National Toxicology Program
NIEHS
National Institutes of Health
P.O. Box 12233
Research Triangle Park, NC 27709
T: 919-541-4532; F: 919-541-4225
bucher@niehs.nih.gov

Dr. Christopher P. Weis
Toxicology Liaison/Senior Advisor
Office of the Director
NIEHS
National Institutes of Health
31 Center Drive, Room B1C02
Bethesda, MD 20892-2256
T: 301-496-3511
christopher.weis@nih.gov

NIH/NIGMS

Dr. Catherine Lewis
Director, Division of Cell Biology and Biophysics
National Institute of General Medical Sciences
National Institutes of Health
45 Center Drive, Rm. MSC 6200
Bethesda, MD 20892-6200
T: 301-594-0828; F: 301-480-2004
catherine.lewis@nih.hhs.gov

NIOSH

Dr. Charles L. Geraci, Jr.
Coordinator, Nanotechnology Research Center
National Institute for Occupational Safety and Health
4676 Columbia Parkway
Cincinnati, OH 45226
T: 513-533-8339
CGeraci@cdc.gov

Dr. Vladimir V. Murashov
Special Assistant to the Director
National Institute for Occupational Safety and Health
OD/NIOSH M/S-P12
395 E Street, S.W.
Suite 9200
Washington, DC 20201
T: 202-245-0668; F: 202-245-0628
vladimir.murashov@cdc.hhs.gov

NIST

Dr. Heather M. Evans
Senior Program Analyst
Program Coordination Office
National Institute of Standards and Technology
100 Bureau Drive, Stop 1060
Gaithersburg, MD 20899-1060
T: 301-975-4525; F: 301-216-0529
Heather.evans@nist.gov

Dr. Ajit Jillavenkatesa
Senior Standards Policy Adviser
Standards Coordination Office
National Institute of Standards and Technology
100 Bureau Drive, Stop 1060
Gaithersburg, MD 20899-1060
T: 301-975-8519; F: 301-975-4715
ajit.jilla@nist.gov

Dr. Debra L. Kaiser
Technical Program Director
Materials Measurement Science Division
National Institute of Standards and Technology
100 Bureau Drive, Stop 8520
Gaithersburg, MD 20899-8520
T: 301-975-6759; F: 301-975-5334
debra.kaiser@nist.gov

NRC

Mr. Brian Thomas
Deputy Director
Division of Engineering
Office of Nuclear Regulatory Research
Nuclear Regulatory Commission
Washington, DC 20555-0001
T: 301-251-7448; F:301-251-7425
brian.thomas@nrc.gov

NSF

Dr. Parag R. Chitnis
Deputy Director, Directorate for Biological Sciences
National Science Foundation
4201 Wilson Blvd., Suite 605N
Arlington, VA 22230
T: 703-292-8440; F: 703-292-9061
pchitnis@nsf.gov

Dr. Khershed Cooper
Program Director, Nanomanufacturing
National Science Foundation
Directorate for Engineering
4201 Wilson Blvd., Suite 505
Arlington, VA 22230

T: 703-292-8301; F: 703-292-9013
khcooper@nsf.gov

Dr. Barbara Karn
Program Director, Division of Chemical, Bioengineering, Environmental, and Transport Systems
National Science Foundation
Directorate for Engineering
4201 Wilson Blvd., Suite 565 S
Arlington, VA 22230
T: 703-292-7949, F: 703-292-9054
bkarn@nsf.gov

Dr. Fred Kronz
Program Director, Division of Science, Technology, and Society
National Science Foundation
Directorate for Social, Behavioral, and Economic Sciences
4201 Wilson Blvd., Suite 995 N
Arlington, VA 22230
T: 703-292-7283
fkronz@nsf.gov

Dr. Thomas Rieker
Program Director, Division of Materials Research
National Science Foundation
Directorate for Mathematical and Physical Sciences
4201 Wilson Blvd., Suite 1065
Arlington, VA 22230
T: 703-292-4914, F: 703-292-9035
trieker@nsf.gov

Dr. Mihail C. Roco
Senior Advisor for Nanotechnology
National Science Foundation
Directorate for Engineering
4201 Wilson Blvd., Suite 505; Arlington, VA 22230
T: 703-292-8301; F: 703-292-9013
mroco@nsf.gov
Dr. Grace J. Wang
Division Director, SBIR/STTR
Division of Industrial Innovation and

Partnerships
National Science Foundation
4201 Wilson Blvd., Rm. 590
Arlington, VA 22230
T: 703-292-2214; F: 703-292-9057
jiwang@nsf.gov

USDA/ARS

Dr. Robert Fireovid
National Program Leader, Biobased
Products and Biorefining
USDA Agricultural Research Service
5601 Sunnyside Avenue
GWCC 4-2264
Beltsville, MD 20705-5140
T: 301-504-4774; F: 301-504-6231
Robert.Fireovid@ars.usda.gov

USDA/NIFA

Dr. Hongda Chen
National Program Leader, Bioprocess
Engineering/Nanotechnology
National Institute of Food and Agriculture
U.S. Department of Agriculture
1400 Independence Ave., SW
Mail Stop 2220
Washington, DC 20250-2220
T: 202-401-6497; F: 202-401-4888
HCHEN@nifa.usda.gov

USDA/FS

Dr. World Nieh
National Program Leader, Forest Products
and Utilization
USDA Forest Service
1400 Independence Ave., SW
Mailstop Code: 1114
Washington, DC 20250-1114
T: 703-605-4197; F: 703-605-5137
wnieh@fs.fed.us

Dr. Theodore H. Wegner
Assistant Director
Forest Products Laboratory
USDA Forest Service
One Gifford Pinchot Drive

Madison, WI 53726-2398
T: 608-231-9434; F: 608-231-9567
twegner@fs.fed.us

USGS

Dr. Sarah Gerould
Staff Scientist
Ecosystems Mission Area
U.S. Geological Survey
Mail Stop 301 National Center
12201 Sunrise Valley Drive
Reston, VA 20192
T: 703-648-6895; F: 703-648-4238
sgerould@usgs.gov

Dr. Herbert Buxton
Coordinator, Toxic Substances Hydrology
Program
Environmental Health Mission Area
U.S. Geological Survey
3450 Princeton Pike
Lawrenceville, NJ 08648
T: 609-771-3944; F: 609-203-5853
hbuxton@usgs.gov

USITC

Ms. Elizabeth R. Nesbitt
International Trade Analyst for Biotechnology
and Nanotechnology
Chemicals and Textiles Division
Office of Industries
U.S. International Trade Commission
500 E Street, S.W.
Washington, DC 20436
T: 202-205-3355; F: 202-205-3161
elizabeth.nesbitt@usitc.gov

USPTO

Mr. David R. Gerk
Patent Attorney
Office of Policy and External Affairs
Patent and Trademark Office
P.O. Box 1450
Alexandria, VA 22313
T: 571-272-9300; F: 571-273-0085
david.gerk@uspto.gov

Mr. Bruce Kisliuk
Deputy Commissioner for Patent
Administration
Patent and Trademark Office
P.O. Box 1450
Alexandria, VA 22313
T: 571-272-8800
Bruce.Kisliuk@uspto.gov

Ms. Gladys Corcoran
Technology Center Group Director
Patent and Trademark Office
P.O. Box 1450
Alexandria, VA 22313
T: 571-272-1300
Gladys.Corcoran@uspto.gov

Additional Staff Contacts,
National Nanotechnology Coordination
Office (NNCO)

Dr. Tarek R. Fadel
Staff Scientist
NSET and NEHI Executive Secretary
NNCO
4201 Wilson Blvd.
Stafford II, Suite 405
Arlington, VA 22230
T: 703-292-7926; F: 703-292-9312
tfadel@nnco.nano.gov

Dr. Lisa E. Friedersdorf
Sr. Scientist, NSI Program Manager
NNCO
4201 Wilson Blvd.
Stafford II, Suite 405
Arlington, VA 22230
T: 703-292-4533; F: 703-292-9312
lfriedersdorf@nnco.nano.gov

Mr. Geoffrey M. Holdridge
Policy Analyst and Contract Staff Director
NNCO
4201 Wilson Blvd. Stafford II, Suite 405
Arlington, VA 22230
T: 703-292-4532; F: 703-292-9312
gholdrid@nnco.nano.gov

Dr. Mike Kiley
Industry and State Liaison
NILI Executive Secretary
NNCO
4201 Wilson Blvd.
Stafford II, Suite 405
Arlington, VA 22230
T: 703-292-4399; F: 703-292-9312
mkiley@nnco.nano.gov

Ms. Marlowe Newman
Communications Director
NNCO
4201 Wilson Blvd.
Stafford II, Suite 405
Arlington, VA 22230
T: 703-292-7128; F: 703-292-9312
mnewman@nnco.nano.gov

Dr. Stacey Standridge
Policy Analyst
NNCO
4201 Wilson Blvd.
Stafford II, Suite 405
Arlington, VA 22230
T: 703-292-8103; F: 703-292-9312
sstrandridge@nnco.nano.gov

End Notes

[1] General note: In conformance with Office of Management and Budget style, references to years in this report are to fiscal years unless otherwise noted.

[2] Participants in the NNI include Federal departments, independent agencies, and independent commissions, which are collectively referred to as participating "agencies" within this document.

[3] *The National Nanotechnology Initiative Strategic Plan* (NSTC, Washington DC, 2014; www.nano.gov /2014StrategicPlan), p. 5.

[4] *National Nanotechnology Initiative Strategic Plan* (NSTC, Washington DC, 2014, www.nano.gov/2014Strategic Plan), p. 39.

[5] www.WhiteHouse.gov/sites/default/files/microsites/ostp/fy 15 memo m-13-16.pdf.

[6] *The National Nanotechnology Initiative Environmental, Health, and Safety Research Strategy* (NSTC, Washington DC, 2011; www.nano.gov/2011EHSStrategy).

[7] www.WhiteHouse.gov/sites/default/files/microsites/ostp/fy15memom-13-16.pdf

[8] See Table 1 or Appendix B for explanations of the agency abbreviations used on this page and throughout the remainder of this report.

[9] See www.nano.gov/2013BudgetSupplement.

[10] A significant portion of SBIR/STTR funding reported by DOE for 2009 and 2010 is from the American Recovery and Reinvestment Act (ARRA) of 2009.

[11] SBIR/STTR award levels for USDA include FS.

[12] www.nano.gov/2014StrategicPlan

[13] Details for Department of Defense investments are provided per the statutory requirement for DOD reporting on its nanotechnology investments (10 USC §2358). Additional details on DOD investments can be found in Appendix A.

[14] www.cdc.gov/niosh/docs/2014-106

[15] www.usda.gov/wps/portal/usda/usdamediafb?contentid=2013/12/0235.xml&printable=true&contentidonly=true

[16] Details on NNI strategic goals and objectives can be found in the 2014 *National Nanotechnology Initiative Strategic Plan,* www.nano.gov/2014StrategicPlan or www.nano.gov/about-nni/what/vision-goals.

[17] For additional information on the Nanotechnology Signature Initiatives, see the 2014 NNI Strategic Plan www.nano.gov/2014StrategicPlan and www.nano.gov/signatureinitiatives.

[18] Discussed above under coordinated activities in support of Objective 1.2 (p. 28).

[19] www.techconnectworld.com/Nanotech2013/about

[20] www.techconnectworld.com/Nanotech2013/about

[21] www.techconnectworld.com/Nanotech2013/about

[22] www.nano.gov/DRL

[23] www.techconnectworld.com/Nanotech2013/about

[24] For additional information on the NSI topic on nanosensors, including its thrust areas, see www.nano.gov/NSISensors.

[25] See nimss.umd.edu/lgu v2/homepages/home.cfm?trackID=13116

[26] www.techconnectworld.com/Nanotech2013/about

[27] www.federalregister.gov/a/2013-23916

[28] www.usnanosurvey.org

[29] Scott E. McNeil (Ed.), *Characterization of nanoparticles intended for drug delivery*. Springer/Humana Press, 2011.

[30] For the full descriptions of the objectives of this NNI goal, see the 2014 National Nanotechnology Initiative Strategic Plan, www.nano.gov/2014StrategicPlan or www.nano.gov/about-nni/what/vision-goals.

[31] For more information on NNI-supported research infrastructure see www.nano.gov/centers-networks.

[32] See www.nisenet.org/nanodays.

[33] www.nano.gov/you/environmental-health-safety

[34] www.cdc.gov/niosh/topics/PtD/nanoworkshop/default.html

[35] www.cdc.gov/niosh/topics/nanotech/

[36] www.cdc.gov/niosh/docs/2014-102/pdfs/2014-102.pdf

[37] www.nano.gov/2014BudgetSupplement

[38] National Research Council. *Triennial review of the National Nanotechnology Initiative*. Washington DC: The National Academies Press, 2013 (www.nap.edu/catalog.php?record id=18271).

[39] National Research Council. *Research progress on environmental, health, and safety aspects of engineered nanomaterials*. Washington DC: The National Academies Press, 2013(www.nap.edu/catalog.php?record id=18475).

[40] National Research Council. *A research strategy for environmental, health, and safety aspects of engineered nanomaterials*. Washington DC: The National Academies Press, 2012.

[41] Details for Department of Defense investments are provided per the statutory requirement for DOD reporting on its nanotechnology investments (10 USC §2358).

[42] Additional NNCO staff contact information is provided on p. 78.

INDEX

C

D

E

F

G

H

L

M

N

O

Q

R

S

T

U

V